Reviews of Environmental Contamination and Toxicology

VOLUME 226

For further volumes:
http://www.springer.com/series/398

Reviews of Environmental Contamination and Toxicology

Editor
David M. Whitacre

VOLUME 226

Springer

Coordinating Board of Editors

Please note that additional material for this book can be downloaded from
http://extras.springer.com

ISSN 0179-5953
ISBN 978-1-4899-8704-4   ISBN 978-1-4614-6898-1 (eBook)
DOI 10.1007/978-1-4614-6898-1
Springer New York Heidelberg Dordrecht London

Foreword

International concern in scientific, industrial, and governmental communities over traces of xenobiotics in foods and in both abiotic and biotic environments has justified the present triumvirate of specialized publications in this field: comprehensive reviews, rapidly published research papers and progress reports, and archival documentations. These three international publications are integrated and scheduled to provide the coherency essential for nonduplicative and current progress in a field as dynamic and complex as environmental contamination and toxicology. This series is reserved exclusively for the diversified literature on "toxic" chemicals in our food, our feeds, our homes, recreational and working surroundings, our domestic animals, our wildlife, and ourselves. Tremendous efforts worldwide have been mobilized to evaluate the nature, presence, magnitude, fate, and toxicology of the chemicals loosed upon the Earth. Among the sequelae of this broad new emphasis is an undeniable need for an articulated set of authoritative publications, where one can find the latest important world literature produced by these emerging areas of science together with documentation of pertinent ancillary legislation.

Research directors and legislative or administrative advisers do not have the time to scan the escalating number of technical publications that may contain articles important to current responsibility. Rather, these individuals need the background provided by detailed reviews and the assurance that the latest information is made available to them, all with minimal literature searching. Similarly, the scientist assigned or attracted to a new problem is required to glean all literature pertinent to the task, to publish new developments or important new experimental details quickly, to inform others of findings that might alter their own efforts, and eventually to publish all his/her supporting data and conclusions for archival purposes.

In the fields of environmental contamination and toxicology, the sum of these concerns and responsibilities is decisively addressed by the uniform, encompassing, and timely publication format of the Springer triumvirate:

Reviews of Environmental Contamination and Toxicology [Vol. 1 through 97 (1962–1986) as Residue Reviews] for detailed review articles concerned with any aspects of chemical contaminants, including pesticides, in the total environment with toxicological considerations and consequences.

Bulletin of Environmental Contamination and Toxicology (Vol. 1 in 1966) for rapid publication of short reports of significant advances and discoveries in the fields of air, soil, water, and food contamination and pollution as well as methodology and other disciplines concerned with the introduction, presence, and effects of toxicants in the total environment.

Archives of Environmental Contamination and Toxicology (Vol. 1 in 1973) for important complete articles emphasizing and describing original experimental or theoretical research work pertaining to the scientific aspects of chemical contaminants in the environment.

Manuscripts for Reviews and the Archives are in identical formats and are peer reviewed by scientists in the field for adequacy and value; manuscripts for the *Bulletin* are also reviewed, but are published by photo-offset from camera-ready copy to provide the latest results with minimum delay. The individual editors of these three publications comprise the joint Coordinating Board of Editors with referral within the board of manuscripts submitted to one publication but deemed by major emphasis or length more suitable for one of the others.

Coordinating Board of Editors

Preface

The role of *Reviews* is to publish detailed scientific review articles on all aspects of environmental contamination and associated toxicological consequences. Such articles facilitate the often complex task of accessing and interpreting cogent scientific data within the confines of one or more closely related research fields.

In the nearly 50 years since *Reviews of Environmental Contamination and Toxicology* (formerly *Residue Reviews*) was first published, the number, scope, and complexity of environmental pollution incidents have grown unabated. During this entire period, the emphasis has been on publishing articles that address the presence and toxicity of environmental contaminants. New research is published each year on a myriad of environmental pollution issues facing people worldwide. This fact, and the routine discovery and reporting of new environmental contamination cases, creates an increasingly important function for *Reviews*.

The staggering volume of scientific literature demands remedy by which data can be synthesized and made available to readers in an abridged form. *Reviews* addresses this need and provides detailed reviews worldwide to key scientists and science or policy administrators, whether employed by government, universities, or the private sector.

There is a panoply of environmental issues and concerns on which many scientists have focused their research in past years. The scope of this list is quitebroad, encompassing environmental events globally that affect marine and terrestrial ecosystems; biotic and abiotic environments; impacts on plants, humans, and wildlife; and pollutants, both chemical and radioactive; as well as the ravages of environmental disease in virtually all environmental media (soil, water, air). New or enhanced safety and environmental concerns have emerged in the last decade to be added to incidents covered by the media, studied by scientists, and addressed by governmental and private institutions. Among these are events so striking thatthey are creating a paradigm shift. Two in particular are at the center of ever increasing media as well as scientific attention: bioterrorism and global warming. Unfortunately, these very worrisome issues are now superimposed on the already extensive list of ongoing environmental challenges.

The ultimate role of publishing scientific research is to enhance understanding of the environment in ways that allow the public to be better informed. Theterm "informed public" as used by Thomas Jefferson in the age of enlightenment conveyed the thought of soundness and good judgment. In the modern sense, being "well informed" has the narrower meaning of having access to sufficient information. Because the public still gets most of its information on science and technology from TV news and reports, the role for scientists as interpreters and brokers of scientific information to the public will grow rather than diminish. Environmentalismis the newest global political force, resulting in the emergence of multinational consortiato control pollution and the evolution of the environmental ethic. Will the new politics of the twenty-first century involve a consortium of technologists and environmentalists, or a progressive confrontation? These matters are of genuine concern to governmental agencies and legislative bodies around the world.

For those who make the decisions about how our planet is managed, there is an ongoing need for continual surveillance and intelligent controls to avoid endangering the environment, public health, and wildlife. Ensuring safety-in-use of the many chemicals involved in our highly industrialized culture is a dynamic challenge, forthe old, established materials are continually being displaced by newly developed molecules more acceptable to federal and state regulatory agencies, public healthofficials, and environmentalists.

Reviews publishes synoptic articles designed to treat the presence, fate, and, if possible, the safety of xenobiotics in any segment of the environment. These review scan be either general or specific, but properly lie in the domains of analytical chemistry and its methodology, biochemistry, human and animal medicine, legislation, pharmacology, physiology, toxicology, and regulation. Certain affairs in food technology concerned specifically with pesticide and other food-additive problems may also be appropriate.

Because manuscripts are published in the order in which they are received in final form, it may seem that some important aspects have been neglected at times. However, these apparent omissions are recognized, and pertinent manuscripts are likely in preparation or planned. The field is so very large and the interests in it are so varied that the editor and the editorial board earnestly solicit authors and suggestions of under represented topics to make this international book series yet more useful and worth while.

Justification for the preparation of any review for this book series is that it deals with some aspect of the many real problems arising from the presence of foreign chemicals in our surroundings. Thus, manuscripts may encompass case studies from any country. Food additives, including pesticides, or their metabolites that may persist into human food and animal feeds are within this scope. Additionally, chemical contamination in any manner of air, water, soil, or plant or animal life is within these objectives and their purview.

Manuscripts are often contributed by invitation. However, nominations for new topics or topics in areas that are rapidly advancing are welcome. Preliminary communication with the editor is recommended before volunteered review manuscripts are submitted.

Summerfield, NC, USA   David M. Whitacre

Contents

Current Approaches for Mitigating Acid Mine Drainage

Prafulla Kumar Sahoo, Kangjoo Kim, Sk. Md. Equeenuddin, and Michael A. Powell

Contents

P.K. Sahoo (✉)
Department of Environmental Engineering, Kunsan National University,
Jeonbuk 573-701, Republic of Korea

Present Address: Vale Institute of Technology - Sustainable Development,
Rua Boa Ventura da Silva, 955, Nazaré, 66055-090 Belém, Pará, Brazil
e-mail: prafulla.iitkgp@gmail.com

K. Kim (✉)
Department of Environmental Engineering, Kunsan National University,
Jeonbuk 573-701, Republic of Korea
e-mail: kangjoo@kunsan.ac.kr

Sk.Md. Equeenuddin
Department of Mining Engineering, National Institute of Technology,
Rourkela 769008, Odisha, India

M.A. Powell
Faculty of Agriculture, Life and Environmental Sciences, Department of Renewable Resources,
University of Alberta, Edmonton, AB, Canada T6G 2R3

D.M. Whitacre (ed.), *Reviews of Environmental Contamination and Toxicology*
Volume 226, Reviews of Environmental Contamination and Toxicology 226,
DOI 10.1007/978-1-4614-6898-1_1, © Springer Science+Business Media New York 2013

1   Introduction

Acid mine drainage (AMD) is one of the largest environmental problems faced by mining and mineral industries globally (Akcil and Koldas 2006; Akabzaa et al. 2007). AMD is mainly associated with sulphide-rich metalliferous ore deposits, viz., copper, lead, zinc, gold, nickel, tin, and uranium mines (Naicker et al. 2003; Nieto et al. 2007) and coal mines (Bell et al. 2001; Sahoo et al. 2012), which were formed under marine conditions and contain abundant reactive framboidal pyrite (Carrucio and Fern 1974; Campbell et al. 2001). In Table 1, we present a list of well-known AMD sites worldwide.

Worldwide, several sites face AMD problems. A famous massive sulfide deposit exists at Iron Mountain in California. This site has attracted negative worldwide attention for discharging low-pH mine waste having the highest metal concentrations (Nordstrom et al. 2000). Ria of Huelva in Southwestern Spain is one of the most heavily metal-contaminated estuaries in the world as a result of AMD derived from the Iberian pyrite belt (Sáinz et al. 2004). In Germany, there are more than 100 acid lakes that have a pH <4 as a consequence of sulfide mineral oxidation (Geller et al. 1998). In Korea, there are 41 coal mines that together discharge a total AMD waste of more than 141,000 m^3/day (Chon and Hwang 2000). In the USA, Pennsylvania has a long history of coal mining, and AMD constitutes this state's single greatest source of water pollution, being responsible for over 2,400 miles of polluted streams (Mallo 2011). Kleinmann (1989) estimated that nearly 19,300 km of streams and more than 180,000 acres of lakes and reservoirs in the USA have been contaminated by AMD.

AMD is predominantly caused when sulfide minerals that are present in metallic ores, coal beds or in strata overlying and underlying the coal are exposed to oxidation (Jennings et al. 2000; Montero et al. 2005; Lottermoser 2007). In Table 2, we show the sulfide minerals that are responsible for AMD. Pyrite (FeS_2) is the most common precursor of AMD, partially because it is the major sulfide mineral in coal and ores, as well as in the Earth's crust (Egiebor and Oni 2007; Lindgren et al. 2011). The major chemical reactions involved in pyrite oxidation (Kleinmann et al. 1981) are shown in Fig. 1.

Pyrite, when exposed to air and water, releases water soluble components such as Fe^{2+}, SO_4^{2-} and H^+ (see reaction 1 in Fig. 1). The Fe^{2+} produced is oxidized into Fe^{3+} according to reaction [2] (Fig. 1). This reaction is the "rate limiting" step as it is pH dependent; Fe^{3+} in turn either hydrolyzes into amorphous Fe-oxides [3] (Fig. 1) or oxidizes pyrite [4] (Fig. 1), which is accompanied by the release of additional amounts of acids (Nordstrom 1982). At neutral to alkaline pH, the rate of Fe^{2+} oxidation increases and the Fe^{3+} concentration decreases rapidly from precipitation of Fe-oxides [3]. When Fe^{3+} oxides precipitate, atmospheric O_2 is the important oxidant of pyrite. At low pH (<3.5), the hydrolysis of Fe^{3+} ion stops and the activity of Fe^{2+} in solution increases (Zhuping et al. 1998). Further, Fe-oxidizing bacteria such as *Thiobacillus ferrooxidans* catalyze the rate of Fe^{2+} oxidation by a factor of 10^6 times over the rate that occurs by atmospheric O_2 alone (Singer and Stumm 1970).

Table 1 Well-known coal and metalliferous sulfide deposits that exist around the world and are associated with AMD

Deposit	Location	References
Coal	Pennsylvania Coalfield, USA	Cravotta et al. (2010)
	Witbank Coalfield, South Africa	Bell et al. (2001)
	Wangaloa Coal mines, New Zealand	Black and Craw (2001)
	South Coalfield, Brazil	Lattuada et al. (2009)
	Okpara Coalmine, Nigeria	Nganje et al. (2010)
	Makum and Jantia Coalfield, India	Equeenuddin et al. (2010); Sahoo et al. (2012)
	Guizhou Coalmine, China	Tao et al. (2012)
Copper	Richmond Mines, Iron Mountain, USA	Nordstrom et al. (2000)
	Britannia Mines, Canada	Grout and Levings (2001)
Copper-lead-zinc	Iberian Pyrite Belt, Spain	Nieto et al. (2007)
Lead-zinc	Picher mines, USA	Sheibach et al. (1982)
	Carnoules mine, France	Casiot et al. (2003a, b)
Gold	Witwatersrand, Johannesburg, South Africa	Naicker et al. (2003)
	Hillside Mine, Western Arizona, USA	Rampe and Runnells (1989)
Uranium	Curilo, Western Bulgaria	Groudev et al. (2008)
	Poços de Caldas, Brazil	Fernandes and Franklin (2001)
Silver	Montalbion Mine, North Queensland	Harris et al. (2003)
Tin	Ervedosa Mine, Portugal	Gomes and Favas (2006)
Nickel	Black Swan, Western Australia	Lei et al. (2010)

Table 2 The major minerals that are associated with AMD (Jennings et al. 2000; Montero et al. 2005; Lottermoser 2007)

Minerals	Composition
Pyrite	FeS_2
Pyrrhotite	$Fe1-xS$
Marcasite	FeS_2
Arsenopyrite	$FeS_2 \cdot FeAs$
Bornite	$CuFeS_4$
Chalcopyrite	$CuFeS_2$
Galena	PbS
Sphalerite	ZnS
Millerite	NiS
Covellite	CuS

Therefore, under such conditions, Fe^{3+}, rather than O_2, plays the major role in the oxidation of pyrite and corresponding genesis of AMD (Singer and Stumm 1970).

The acidic leachate resulting from pyrite oxidation further reacts with rocks and may contribute significant amounts of SO_4^{2-} and toxic metals to the acidic drainage, which is responsible for the environmental deterioration of stream-, lake-, ground-water and soils (Ji et al. 2007; Pelo et al. 2009; Equeenuddin et al. 2010, 2013; Sahoo et al. 2012). Some of the more dangerous effects of AMD include fish kills, loss of other aquatic species, effects on vegetation (Kargbo et al. 1993; Bell et al. 2001; Nordstrom and Alpers 1999) and corrosion of mining equipment or structures such as barges, bridges, and concrete emplacements (Goel 2006).

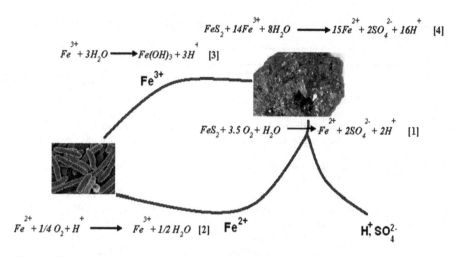

$$FeS_2 + 14Fe^{3+} + 8H_2O \longrightarrow 15Fe^{2+} + 2SO_4^{2-} + 16H^+ \quad [4]$$

$$Fe^{3+} + 3H_2O \longrightarrow Fe(OH)_3 + 3H^+ \quad [3]$$

$$Fe^{3+}$$

$$FeS_2 + 3.5\,O_2 + H_2O \longrightarrow Fe^{2+} + 2SO_4^{2-} + 2H^+ \quad [1]$$

$$Fe^{2+} + 1/4\,O_2 + H^+ \longrightarrow Fe^{3+} + 1/2\,H_2O \quad [2] \quad Fe^{2+} \qquad H^+,SO_4^{2-}$$

Fig. 1 A simplified diagram showing the chemical process by which pyrite is oxidized

Numerous strategies have been proposed to control AMD (Johnson and Hallberg 2005; Hurst et al. 2002; Kleinmann 1990; Gazea et al. 1996), most common of which involve postproduction treatments, in which neutralization and precipitation techniques with alkaline materials are utilized. However, such treatments are often unsuitable because they are difficult to manage, require large amounts of alkaline materials and generate huge waste by-products, which require significant investment, and may produce long-term undesirable effects (Sharma 2010). A more viable option for reducing AMD is to prevent sulfide oxidation at the source (Filion et al. 1990; Evangelou 1995a). If achieved, prevention can provide a permanent solution and is low cost (Nyavor and Egiebor 1995). There have been several strategies developed to prevent pyrite oxidation and subsequent formation of AMD (Evangelou 1995a; Brown et al. 2002; Zhang and Evangelou 1996; Johnson and Hallberg 2005). These methods are classified into five major types: physical barriers, bacterial inhibition, chemical barriers (or passivation), electrochemical protection, and desulfurization (Fig. 2).

Physical barriers create an oxygen barrier that isolates pyrite from weathering reactants (Belzile et al. 1997; Peppas et al. 2000; Vigneault et al. 2001). Bacterial inhibition suppresses the activities of Fe-oxidizing bacteria (Kleinmann et al. 1981; Lalvani et al. 1990; Kleinmann 1990). During chemical passivation, sometimes called microencapsulation, a protective organic or inorganic coating is formed on pyrite surfaces (Lalvani et al. 1996; Nyavor et al. 1996; Zhang and Evangelou 1998); for the electrochemical process, dissolved O_2 is removed from sulfide tailings by using electrochemical principles (ENPAR Technologies Inc 2000). In the desulfurization process, sulfide-rich minerals are separated from the tailings by using a flotation technique, leaving behind the majority of low-sulfur content tailings (Bois et al. 2004).

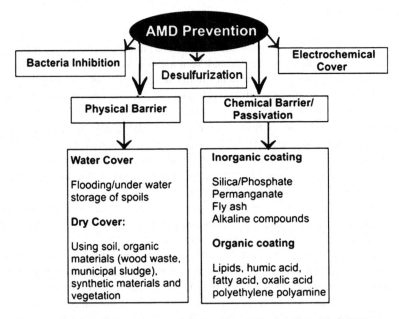

Fig. 2 Various approaches that have been used to prevent acid mine drainage (AMD)

In this paper, we present a comprehensive review of the various techniques that are available to prevent formation of AMD, and compare these techniques for their long-term effectiveness, cost, and impacts. In addition, we highlight emerging technologies that may be used to address AMD. Insights from this paper may help researchers and environmental engineers to select suitable methods for addressing site-specific AMD problems.

2 Prevention of AMD

The primary aim of preventing AMD is to minimize the supply of the oxidants responsible for sulfide oxidation. Below, we describe several strategies to prevent formation of AMD from sulfide-bearing wastes.

2.1 Physical Barriers

The most common and traditional approach for controlling AMD is to establish a physical barrier that comprises a wet or dry cover (Kleinmann 1990; Evangelou and Zhang 1995). Such dry or wet covers work because they create an oxygen barrier that prevents pyrite oxidation (Belzile et al. 1997; Peppas et al. 2000; Vigneault et al. 2001).

2.1.1 Water Cover

Disposal of sulfide-bearing wastes under water is a suitable technique to limit acid generation (Davé et al. 1997; Yanful and Verma 1999; Simms et al. 2000; Demers et al. 2008). By disposing of such wastes under water, highly anoxic conditions are created in tailings and inhibit sulfide oxidation. In addition, microbial catalysis, in association with other mechanisms such as metal hydroxide precipitation and development of sediment barriers between tailing and overlying waters also inhibit oxidation (Kleinmann and Crerar 1979). The faster a mine floods, the less time there is for the pyrite to oxidize. Both laboratory and field studies have shown that underwater disposal is one of the most effective ways to stabilize sulfide tailings (Pedersen et al. 1991, 1993; Yanful et al. 2000; MEND 2001).

Large-scale mines typically use water cover to reduce the rate of oxygen contact with the sulfide-bearing wastes since it is comparatively less expensive than other options. However, using a water cover may not achieve the long-term purpose. Vigneault et al. (2001) reported that there was clear evidence of sulfide oxidation and mobilization of Cd and Zn after 2 years of water cover of sulfidic tailings. Furthermore, a water cover for surface-mined lands may not be a practical alternative. Lapakko (1994) and Dave and Vivyurka (1994) both indicate that subaqueous disposal of sulfide tailing has only a minor effect on reducing sulfide oxidation rates. Moreover, this technique is limited to sites that can be flooded, or where the water table may be permanently altered to cover sulfide wastes, or where waste materials can be placed in a lake. Johnson and Hallberg (2005) reported that water cover is problematic at sites where the influx of oxygen-containing water occurs, or where mines are only partially flooded. This is due to rise and fall of the water table and seasonal changes, which allow pyrite to oxidize when exposed to the atmosphere. Another problem arises if the contaminant storage system is too leaky, which is the case in Canada, where more than half of the sulfide mine-waste sites are not physically suited for utilizing water cover (Wang et al. 2006). Finally, flooding of mine wastes requires a rigorous engineering design and proper maintenance to minimize the risk of dike failure, which is often not cost effective.

2.1.2 Dry Cover

Dry cover primarily comprises a sealing layer with low hydraulic conductivity (Fig. 3), which restricts the oxygen supply and limits water percolation into the tailings, thereby reducing the rate of pyrite oxidation (Kleinmann 1990; Hallberg et al. 2005). Materials used for dry cover may be composed of fine-grained soil or a soil substitute, organic material, synthetic material or vegetation (Yanful et al. 2000; Holmström et al. 2001; Harrington 2001; Bussiere et al. 2004). In a laboratory-based experiment, Yanful and Payant (1992) demonstrated that soil cover significantly decreased acid generation by three-orders of magnitude, relative to uncovered tailings. However, in a 3-year field study using soil to isolate pyritic mine waste, Yanful and Orlandea (2000) found that the soil cover did not work well as expected, probably due to sidewall passage of oxygen and water. In other studies, the

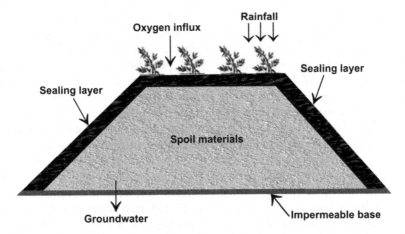

Fig. 3 Diagram of a typical dry cover designed to minimize the production of acidic effluents from sulfide-bearing wastes

effectiveness of soil cover on the surface of waste rock piles at the Rum Jungle mine site in the Northern Territory, Australia was evaluated (Harries and Ritchie 1985, 1987; Timms and Bennett 2000). The performance of the cover was regularly monitored over time, and although the cover reduced the rate of sulfide oxidation initially, oxidation increased with time. Similarly, at Mussel White Mine, Ontario, despite soil cover, the oxidation of pyrite proceeded after a few years from precipitate infiltration (Wang et al. 2006). Yanful (1993) and Yanful et al. (1999) reported a short-term effectiveness and high maintenance cost of soil cover for preventing pyrite oxidation. Swanson et al. (1997) found soil covers to be ineffective in areas that experienced acute wet and dry seasons because during drying cracks developed, which reduce effectiveness relative to that occurred in temperate zones.

In addition to using soil as a dry cover material, synthetic materials such as polyethylene and plastic liners can be used to control AMD under field conditions. Caruccio (1983) used a plastic cover to completely cover a site in West Virginia and found that AMD was significantly reduced. However, covering a large volume of waste with a synthetic liner is usually too expensive, which may preclude its routine field use (Skousen and Foreman 2000). Additionally, plastic or polymer liners are prone to cracking, and repair costs are prohibitive.

Alternative dry cover designs, which reduce pyrite oxidation via various mechanisms (Nicholson et al. 1989; Tasse et al. 1997a), have been used and include organic wastes, such as wood waste, wood chips, municipal sludge, peat, paper-mill sludge, and vegetation. Organic waste provides a pH buffer that neutralizes acids and consumes oxygen, creating anoxic conditions in the tailings, which inhibit Fe-oxidizing bacteria and reduce sulfate to sulfide conversions, thus immobilizing metals. Moreover, as microbial degradation of organic matter proceeds, oxygen levels in the waste are reduced. Reardon and Poscente (1984) evaluated the potential of wood waste to limit oxidation in sulfide-bearing mine tailings and found that it provided an effective oxygen barrier, although the effective life span of this approach was short.

Similarly, Yanful and co-workers (1992, 2000) evaluated the effectiveness of wood bark vs. other potential cover materials, and concluded that it significantly increased the rate of acid production relative to the other covers. The success of wood bark may have resulted from its degradation, which provided nutrients to Fe-oxidizing bacteria that increased their activity. Furthermore, organic matter can lead to the complexation of free Fe(III) by soluble microbial growth products (SMPs) that are produced by microorganisms growing in waste rock. Increases in SMPs reduce the effectiveness of ferric iron as an oxidant. Pandey et al. (2011) performed a lab study on the effect of organic carbon on pyrite oxidation and the role of a secondary mechanism, i.e., complexation of free Fe(III) by SMPs on pyrite oxidation. The results showed a decrease in the rate of pyrite oxidation that was dependent on the concentration of SMPs in solution. However, the mechanism of the SMP complexation and the role of the individual heterotroph in producing SMP are not clear. Organic products also coat pyrite surfaces and minimize reaction area. Backes et al. (1987) studied the oxidation of pyrite in the presence of organic waste materials such as manures and sewage sludge. They found that these organic materials inhibited pyrite oxidation. Although some organic residues may provide a short-term solution to AMD, and may be cost effective, they have other problems. In particular, organic material, derived from pulp and paper production and from municipal sewage sludge, contains organic acids or other agents, e.g., toxic metals, that have the potential to leach vertically in waste areas. Such movement can subsequently contaminate underlying tailings and cause deleterious environmental effects (Watzlaf and Erickson 1986; Tasse et al. 1997a; Peppas et al. 2000; Lollar 2005). Such contamination events may have caused metal pollution in the shallow portion of the Nickel Rim tailings impoundment, near Sudbury, Ontario, Canada (Ribet et al. 1995). Pond et al. (2005) reported that, although biosolid amendment controls acidic drainage and reduces SO_4^{2-} concentration from copper mine tailing, it increases the concentration of toxic metals, such as As, Cu, Ni, and Zn. Using a vegetative cover achieves AMD results that are similar to those provided by other organic materials. Again, however, utilizing vegetation offers only a short-term solution (Silver and Ritcey 1985). For example, Fugill and Sencindiver (1986) reported that establishing vegetation on coal refuses reduced acid generation, but for less than two growing seasons.

2.2   Bacterial Inhibition

Inhibition of Fe- and S-oxidizing bacteria, which actively participate in pyrite oxidation, is another effective way of controlling AMD (Kleinmann and Crerar 1979; Kleinmann 1998; Bacelar-Nicolau and Johnson 1999). Inhibition was achieved by using many bactericides, the most common of which were anionic surfactants (sodium dodecyl sulfate or sodium lauryl sulfate), organic acids, and food preservatives (Kleinmann and Erickson 1981; Dugan and Apel 1983; Backes et al. 1986; Dugan 1987). Bactericides generally alter the protective, greasy coating that allows the internal enzymes of bacteria to maintain a near neutral pH, to function normally in an acid environment, and/or disrupt the contact between the bacteria and the mineral surface (Langworthy 1978; Ingledew 1982; Kleinmann 1998).

Bactericides are often liquid amendments and thus can be easily applied on acid-generating wastes. Bactericides are primarily used in situations where immediate action for control of AMD generation is needed. Many lab-scale studies have been performed on bactericides, including benzoic acid, fatty acids and amines, which were found to be effective in limiting pyrite oxidation in sulfide-bearing mine wastes (Kleinmann et al. 1981; Onysko et al. 1984). The U.S Bureau of Mines conducted a field-based experiment on coal refuse, in which they sprayed 30% sodium lauryl sulfate on a large area containing acid-refusing coal piles (Kleinmann and Erickson 1983). Three months after application, they observed that acidity and sulfate dropped about 60%, while Fe decreased by 90%. However, this treatment was effective for only a few months, because it was easily washed from the waste rocks. Further, it was shown that the application of anionic surfactants are rarely successful at metal mines (Parisi et al. 1994), and that they work best on fresh and unoxidized sulfides (Loos et al. 1989; Johnson and Hallberg 2005). In addition, because bactericides are generally water-soluble and leach from wastes, repetitive treatments are needed to inhibit bacterial populations (Kleinmann 1999). Sand et al. (2007) reported that bactericides did not completely kill sulfide-oxidizing bacteria, but only reduced their activity and numbers. Although bactericides can reduce operating costs compared to dry or wet cover techniques, one crucial limitation to using them is that anionic surfactants are toxic to aquatic organisms (Liwarska et al. 2005; Hodges et al. 2006).

2.3 Microencapsulation

Microencapsulation or (passivation) of sulfide minerals is a new and promising technique among the various source-inhibition technologies for controlling AMD (Evangelou 2001). In passivation, a chemically inert and protective surface coating (inorganic or organic) is created over the sulfide surface, which limits both O_2 and Fe^{3+} attack and inhibits the sulfide oxidation for an extended period (Zhang and Evangelou 1998). We describe various organic and inorganic coatings below, which have been used to achieve passivation.

Inorganic Coatings

Phosphate

Application of phosphate on iron sulfide residues can form a stable iron phosphate coating that physically inhibits access of oxidants and limits further oxidation (Evangelou 1994, 1995a; Georgopoulou et al. 1996). The coating is formed by bringing sulfide residues into contact with a solution containing an oxidant (H_2O_2), buffer, and phosphate salt (KH_2PO_4) (Huang and Evangelou 1994; Evangelou

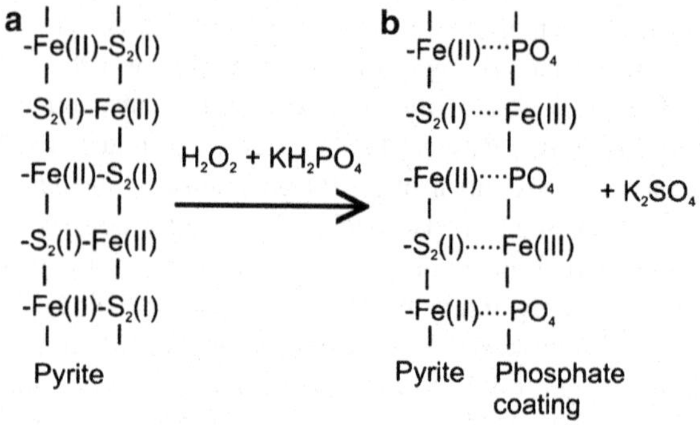

Fig. 4 Schematic of how the phosphate coating on the pyrite surface is formed; *dotted lines* in (**b**) indicate physical bonding between pyrite and FePO$_4$ (Evangelou 1995a)

```
a  |     |                                    b  |     |
  -Fe(II)-S₂(I)                                 -Fe(II)····PO₄
   |     |                                       |     |
  -S₂(I)-Fe(II)                                 -S₂(I) ···· Fe(III)
   |     |              H₂O₂ + KH₂PO₄            |     |
  -Fe(II)-S₂(I)         ───────────────▶        -Fe(II)····PO₄        + K₂SO₄
   |     |                                       |     |
  -S₂(I)-Fe(II)                                 -S₂(I)····Fe(III)
   |     |                                       |     |
  -Fe(II)-S₂(I)                                 -Fe(II)····PO₄
   |     |                                       |     |
    Pyrite                                        Pyrite  Phosphate
                                                          coating
```

1995a, b). In this stepwise process, pyrite is first oxidized in the presence of hydrogen peroxide and releases Fe^{3+}. The Fe^{3+} ion then reacts with PO_4 to form an acidic-resistant precipitate such as $FePO_4$ or $FePO_4 \cdot 2H_2O$ that coats the surface of pyrite. This process is shown schematically in Fig. 4.

Several authors, based on laboratory experiments, have demonstrated that a phosphate coating can significantly inhibit sulfide oxidation and prevent acid generation (Huang and Evangelou 1994; Evangelou 1995b; Nyavor and Egiebor 1995; Georgopoulou et al. 1996; Zhang and Evangelou 1998; Elsetinow et al. 2001; Harris and Lottermoser 2006a; Cárdenes et al. 2009). Recently, Chao et al. (2012) reported that application of KH_2PO_4 significantly reduced sulfide oxidation (reduced sulfate production from 200 to 13 mg/L) from II-Gwang mine samples, in both batch and field experiments. However, this experiment was conducted for only 8 days. Furthermore, as shown by Mauric and Lottermoser (2011), in a field experiment on waste rocks of the Century Pb-Zn Mine, Australia, over the long term, this coating has had only limited success. Evangelou (2001) conducted a long-term field experiment and reported that phosphate treatment on coal waste performed well only for the first 15 weeks, after which oxidation rapidly increased.

Moreover, this coating proved ineffective on polyminerallic mine waste (Harris and Lottermoser 2006b). In field trials, Mauric and Lottermoser (2011) demonstrated that phosphate amendment of metalliferous waste rocks was effective only for the short-term and amendment by phosphate rock or phosphate fertilizer did not improve leachate quality compared to the unamended waste. Also important is the environmental safety aspect; phosphate coatings that enter water causes eutrophication and entails handling H_2O_2 (which is required in the phosphate coating process; see Fig. 4), which may be dangerous (Elsetinow et al. 2003). Phosphate coating was also ineffective in preventing biological oxidation because phosphate is an important nutrient that promotes microbial growth (Lan et al. 2002). Vandiviere and Evangelou (1998) reported that a phosphate coating did not inhibit pyrite oxidation in the presence of sulfide-oxidizing bacteria. Moreover, a phosphate coating is stable only at higher pH levels (>4) Evangelou (1995b); thus, continuous monitoring is required to maintain the desired pH, which is difficult when acidic drainage is present. Further, in such coatings, ferrous phosphate complex may form, which then inactivate the phosphate coating and make the overall process short-lived (Evangelou 2001). However, this

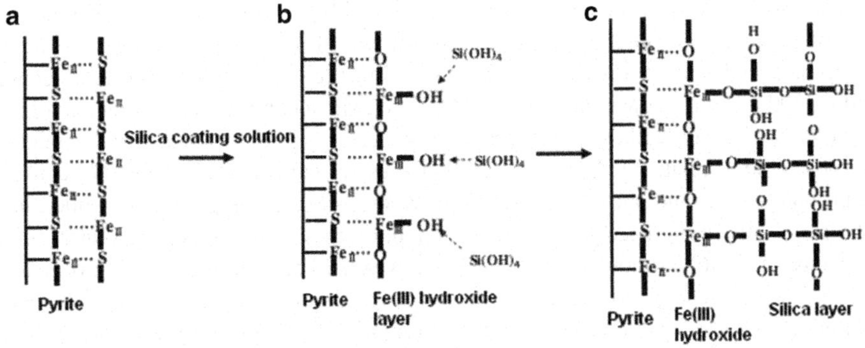

Fig. 5 Diagram of how the silica coating is formed on the pyrite surface; *dotted lines* indicates the interfacial physical contact between pyrite surface and Fe oxide (Bessho et al. 2011)

problem can be effectively managed by combining the phosphate and bactericide techniques. Olson et al. (2005) reported that low concentrations of bactericide, such as thiocyanate are strong and selective inhibitors of microbial iron oxidation, and can inhibit severe acid generation. Such acidic drainage reduces the effectiveness of a phosphate coating on pyrites by fomenting their precipitation. This technique was trialed at the Red Dog zinc-lead mine in northwestern Alaska on sulfidic waste materials, both in lab and field scale (Olson et al. 2005). The results demonstrated that the application of thiocyanate reduced ARD generation by 50% or more, compared to untreated sulfidic waste. Low dosages of phosphate materials treated with thiocyanate reduced ARD generation beyond what was achieved with thiocyanate alone.

Silica

Coating pyrite surfaces with silica was patented by Evangelou (Evangelou 1996) and is a novel technique for reducing oxidation. Using this technique, sulfide-bearing materials are treated with a soluble silica solution containing a buffering agent and a low concentration of an oxidizer such as H_2O_2. The coating is applied as follows: (1) the oxidizer reacts with pyrite and releases ferrous ions, which subsequently is oxidized into ferric ions; (2) in the presence of the buffer solution, the ferric ions form Fe-oxyhydroxide colloids; and (3) silica reacts with the Fe-oxyhydroxide to produce a ferric hydroxide silicate layer on the pyrite surface, and subsequently a silica layer forms on the surface (Fig. 5) (Zhang and Evangelou 1998). Silica coatings seem to be superior since they resist acidic attack (as low as pH 2–4) (Zhang and Evangelou 1998).

Numerous laboratory experiments have shown that a silica coating is significantly effective in preventing sulfide oxidation (Zhang and Evangelou 1998; Takashi et al. 2003; Cárdenes et al. 2009; Kargbo and Chatterjee 2005; Bessho et al. 2011). This process was studied in the field by Vandiviere and Evangelou (1998). These authors conducted an experiment on mine tailings in Canada and found that the silica coating significantly prevented sulfide oxidation. Further, over the long term, coating with silica is more effective than coating with phosphate, because phosphate treatment only delayed the oxidation but did not stop it, whereas silica coatings prevented acid generation over a long period of time (i.e., 6 years) (Eger and Antonson 2002, 2004; Eger and Mitchell 2007). Moreover, silica coatings have attracted a lot of attention because silica is a common component of geologic material and is readily available at relatively low cost. Another plus for silica is that it does not cause eutrophication as the phosphates do. Furthermore, silica coatings can be applied in the field without the need for continuous monitoring and maintaining a neutral pH (Evangelou 1996). Although, thus far, there have only been a limited number of field studies in which silica coatings have been tested, this method presents the first economically feasible and long-term solution to AMD (Evangelou 1996).

Permanganate

Passivation of sulfidic materials by using permanganate was developed by DuPont (De Vries 1996). In this technique, a pyritic rock surface is first rinsed with alkaline solution at a high pH (>12), then is treated with potassium permanganate and magnesium oxide. The overall process creates an inert iron–manganese oxide layer (a coating) on the sulfide surface, which inhibits further oxidation and subsequently reduces acid generation. Magnesium oxide provides an effective coating (Beck 2003), but high pH must be maintained to prevent the permanganate from disproportionating to a weaker oxidant (Thompson and Jerkins 1999).

Permanganate is stable, resistant to acids and less toxic to the environment than many other coatings. De Vries (1996) showed that permanganate passivation is an appropriate option for preventing AMD and that the permanganate can be easily applied in the field without disturbing the target rocks. McCloskey (2005) studied a field-based experiment of permanganate treatment at the Golden Sunlight Mines, located near Whitehall, Montana, wherein caustic solution was first sprayed on the mine wall to adjust the pH to >12, prior to the permanganate application. The results showed that after 41 weeks, acid production, SO_4^{2-} and metals were all reduced. Pilot-scale experiments have revealed that permanganate passivation can significantly reduce contaminant release for more than 5 years, but the long-term effectiveness of this treatment remains to be established. Miller and Van Zyl (2008) reported that this technique has shown the greatest success on freshly mined surfaces, and requires the lowest consumption of reagents compared to aged reactive rock surfaces. Ji et al. (2012) also found less SO_4^{2-} production from Young-Dong coal mine samples in the presence of $KMnO_4$ (16% sulfate production compared to no surface coating agents) within 8 days.

When compared to traditional and alkaline treatments, passivation is a low-cost technique (Kapadia et al. 1999). One of the problems associated with using passivation on weathered rock samples is the large amount of lime required to maintain a pH of 12 (De Vries 1996). Successful application of this technique may be difficult (Glover 2007) and the chemistry of the permanganate coating is still not clear (De Vries 1996; Felipe 2008); hence, more laboratory and long-term field studies are required to better understand the technique and its long term effectiveness.

Alkaline Materials

Addition of alkaline materials to sulfide residues is a relatively common approach for preventing AMD. The most common alkaline materials used include sodium hydroxide (NaOH), sodium carbonate (Na_2CO_3), sodium bicarbonate ($NaHCO_3$), lime (CaO, $Ca(OH)_2$) and limestone ($CaCO_3$) (Brady et al. 1990; Kleinmann 1990; Vandiviere and Evangelou 1998; Evangelou 2001). Alkaline compounds inhibit sulfide oxidation in two ways: (1) they inhibit the activity of oxidizing bacteria (Nicholson et al. 1988), and (2) they neutralize acids, leading to Fe^{3+} precipitation as Fe-hydroxide colloids, which precipitate on the surface of the pyrite (Nicholson et al. 1990; Evangelou 1995a; Huminicki and Rimstidt 2009).

There are several ways that alkaline materials are applied to sulfide residues at mine sites. For underground mines, alkaline material such as lime was applied into mine spoils through boreholes as lime–water slurry. However, this technique was mostly unsuccessful because of slurry settling problems and limited solubility of lime (Kleinmann et al. 1981). This problem was resolved by using more soluble compounds like sodium carbonate and sodium hydroxide, but these replacement chemicals failed to suppress acid generation significantly (Kleinmann et al. 1981). For surface mines, alkaline materials are applied as blends with mine spoil, or alkaline material is placed above or below the mine spoil (Miller et al. 2003, 2006; Mylona et al. 2000; Taylor et al. 2006). Blending can be a difficult task, because it is dependent on the degree of mixing, and the nature of the contact between acidic rock and the alkaline materials. If the mixture is not homogeneous, then acid generation may occur. Common alkaline materials blended with mine spoils include limestone ($CaCO_3$) and lime (CaO or $Ca(OH)_2$).

Limestone is often the least expensive and most readily available source of alkalinity at coal mines. Waddell et al. (1980) used surface application of limestone waste and lime on acid-prone overburden in North Central Pennsylvania. Although it decreased the amount of acid production, the pH increased only to 4.4 from 3.9. Geidel and Caruccio (1984) examined the effects of applying limestone mixed with sulfur compounds to the acidic materials at a reclaimed coal mine. They found that the treatment produced alkaline drainage, but it shortly became acidic. Lusardi and Erickson (1985) used high-Ca crushed limestone to prevent AMD generation at Clarion County, Pennsylvania. However, after 1 year, there was no substantial inhibition of acid generation. Other authors (Ritsema and Groenenberg 1993; Yanful and Orlandea 2000) also noted that this approach did not stop pyrite oxidation for long periods, probably due to the armoring of limestone particles by iron oxyhydroxides (Ritsema and Groenenberg 1993; Al et al. 2000) and the fact that abiotic pyrite oxidation took place (Evangelou 2001; Caldeira et al. 2010). Because these alkaline materials are 'resources' not 'residues', their use can be cost prohibitive, although they are more financially attractive than other methods of passivation.

Recently, fly ash (from coal combustion) was identified as a suitable alternative alkaline material because of its low-cost, local availability, and self-healing capacity (Pérez-López et al. 2007a). Fly ash prevents AMD generation in two ways: (1) over short periods, fly ash increases pH, resulting in the precipitation of Fe oxyhydroxide

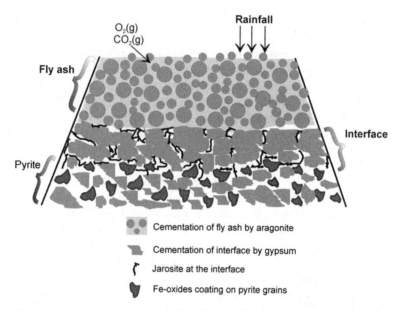

Cementation of fly ash by aragonite

Cementation of interface by gypsum

Jarosite at the interface

Fe-oxides coating on pyrite grains

Fig. 6 Diagram showing the precipitation of neoformed phases at the interface and Fe-oxyhydroxide coatings on pyrite. Redrawn after Pérez-López et al. (2007c)

colloids, which encapsulate the pyrite grains (Fig. 6) over long periods, and fly ash promotes formation of a hardpan at the interface between the fly ash and pyrite from precipitation of neoformed minerals such as gypsum and aragonite (Fig. 6) (Pérez-López et al. 2007b). In addition, the pozzolans in fly ash, such as silica and alumina oxides, produce a calcium silicate gel that binds inert solid particles, i.e., pyrite grains (Palamo et al. 2007). Taken together, these properties produce a strong barrier to oxidants and water diffusion (Blowes et al. 1991; Tasse et al. 1997b). Numerous authors of laboratory studies have reported the use of fly ash for stabilizing pyrite rich tailings and preventing AMD (Xenidis et al. 2002; Pérez-López et al. 2005; Bulusu et al. 2007; Pérez-López et al. 2007b). Field studies from Sudbury Mine, Ontario have also demonstrated that fly ash can effectively stabilize acid-generating mine tailings (Shang et al. 2006). Another study on sulfide mining tailing in Falun, Sweden showed that fly ash significantly decreased the rate of pyrite oxidation (Hallberg et al. 2005). Yeheyis et al. (2009) conducted a column experiment and showed that mixing fly ash with Mussel White Mine tailings effectively controlled AMD over a long period. Lime kiln dust is also used as a suitable alkaline material to prevent AMD. Rich and Hutchison (1994) reported a successful application of lime kiln dust (2%) to coal refuse at a preparation plant in West Virginia.

An added advantage of applying fly ash is that it reduces waste disposal by transforming industrial waste into industrial by-products. Furthermore, use of economically viable and available material makes this technique feasible for the long-term. However, more field-based experiments are needed before it can be recommended for wide spread application. Furthermore, high-carbonate tailings can be used extensively to minimize the risk of acidic leachate from sulfide rich tailing, when using an encapsulation approach. This was recently studied by Mend (2010) at a gold-tailings impoundment site in the Timmins area, Ontario, Canada.

Organic Coatings

Humic Acid

Humic acid is potentially attractive for passivating pyrite (Belzile et al. 1997; Lalvani et al. 1996; Khummalai and Boonamnuayvitaya 2005), because of its high affinity for oxide surfaces (Sposito 1984), resulting in reduced electrochemical activity of pyrite (Duval et al. 2007). Lalvani et al. (1991) found that this method could decrease the rate of pyrite oxidation up to 98%. Work by Ačai et al. (2009) also supports the potential of humic acid for passivating pyrite. Humic acid is generally widely distributed in soil and water (pretreatment for consumption) (Sasaki et al. 1996) and poses no threat to the environment (Ačai et al. 2009). It provides an environmentally suitable alternative to artificial surfactants for cleaning up soil and groundwater contamination from petroleum fuels (Lesage et al. 1995). Humic acid can be sprayed on waste rock piles to form a protective layer, making it a simple and economic passivization method. The disadvantage of humic acid is that it may not be effective for use in highly acidic environments, because humic acid is insoluble at a pH below 2. Furthermore, humic acid's effect on bacterial oxidation and its long-term effectiveness are still questionable; the upshot is that more work is needed to confirm the suitability of humic acid for widespread use.

Lipids

Lipids are used for controlling pyrite oxidation and AMD (Elsetinow et al. 2003). Lipids, containing one or more two-tailed compounds (two hydrophobic tails per hydrophilic), form a hydrophobic barrier that inhibits oxidation (Evangelou and Zhang 1995). Upon hydration, two-tail lipid compounds spontaneously form structures in an aqueous solution that allow the interaction of their polar head with the solution and isolate the hydrophobic tails from the solution (Elsetinow 2006). The use of a single-tail lipid, such as stearic acid, shows no significant effect on pyrite oxidation (Elsetinow 2006). Zhang et al. (2003a) reported that two-tail lipids containing long hydrocarbon tails strongly suppressed pyrite oxidation, even at low pH and under oxic conditions, but no suppression was observed for a short-chain, two-tailed lipid (1,2-dipropionoyl-sn-glycero-3-phosphocholine). This indicates that the difference in suppressed oxidation is attributable to the length of the hydrophobic tail, which has been confirmed in many studies (Elsetinow et al. 2003; Zhang et al. 2003b, 2006; Hao et al. 2006). Lipid coatings were determined to be superior to silica coatings at both pH 2 and 6 (Kargbo et al. 2004).

Recently, Hao et al. (2009) studied the effectiveness of a phospholipid [1,2-bis(10,12-tricosadiynoyl)-sn-glycero-3-phosphocholine (23:2 Diyne PC)], on pyrite oxidation. Pretreatment of pyrite with this phospholipid formed an adsorbed organic layer and reduced the rate of pyrite oxidation in the absence or presence of bacteria (*A. ferrooxidans*). However, after 4–5 days the effectiveness of the lipid in inhibiting pyrite oxidation was lost. In contrast, the lipid layer strongly suppressed pyrite oxidation (up to 30 days) in the presence of both microbial populations when the lipid/pyrite was pretreated with UV radiation, which induces cross-linking of the lipid tails (via polymerization of diacetylene groups in the tails). Lipid coatings are effective under acidic conditions for extended periods in the field, without the need for regular monitoring or adjustment of pH (Elsetinow 2006). Since lipids are mainly composed of carbon, this method is inexpensive and environmentally safe (Elsetinow et al. 2003), and lipids can be safely disposed of without the need for expensive containment. Neither the long-term effect of lipid coatings, nor the formation of lipid assemblages as bilayers on pyrite is well understood, and would benefit from new future research attention.

Polyethylene Polyamine

Polyethylene polyamines prevent sulfide oxidation because of their good buffering capacity and their reducing characteristics (Kirk-Othmer 1984). Cai et al. (2005) showed the inhibitive influence of polyethylene polyamine (e.g., triethylenetetramine, TETA) on oxidation of pyrrhotite. Similarly, Güler (2005) used dithiophosphinate for inhibiting pyrite oxidation. Moreover, Chen et al. (2006) demonstrated the effect of TETA and diethylenetriamine (DETA) on pyrite oxidation and found that both coatings significantly prevented oxidation. TETA was more effective than DETA, because it is more hydrophobic and displays one or more alkyl functional groups. The higher hydrophobic nature of TETA leads to better surface coating and prevents the proliferation of the bacterium *Acidithiobacillus ferrooxidans*, which plays an important role in the oxidation of sulfides (Chen et al. 2006). Despite this effectiveness, these agents may not be suitable for field application because they are toxic to the environment (according to regulation (EC) No. 1272/2008) and are costly.

Alkoxysilanes

Alkoxysilanes generally polymerize upon contact with moisture and produce a strong resistant consolidant (Khummalai and Boonamnuayvitaya 2005). After drying and ageing, they form a stable xerogel with silicon-oxygen composition similar to a stone binder (Mosquera et al. 2002). Khummalai and Boonamnuayvitaya (2005) studied the prevention of arsenopyrite oxidation by using a series of alkoxysilanes, including methyltrimethoxysilane (MTMOS), tetramethoxysilane, tetraethoxysilane, and *N*-(2-aminoethyl)-3-aminopropyl trimethoxysilane. They observed that the MTMOS coating significantly suppressed both biological and chemical oxidation better than the others did. They concluded that the reason the MTMOS performed better was from its crack-free morphology and formation of a hydrophobic $Si-CH_3$ surface group. The formation of the $Si-CH_3$ contributed not only to the hydrophobicity of the surface, but also resulted in the crack-free nature of the coating (Rao et al. 2003). However, this study was conducted at 50 °C, so its effect at typical ambient surface temperatures is largely unknown. Further, reagent costs for this technique are high and, therefore, it may not be suitable for long-term field application.

8-Hydroxyquinoline

Pyrite oxidation can be suppressed by coating the sulfides with a layer of 8-hydroxyquinoline.

Lan et al. (2002) used 0.0034 M 8-hydroxyquinoline in a laboratory experiment and found that it significantly inhibited both chemical and biological pyrite oxidation. However, this approach is likely to be effective only on fresh tailing piles before any acidic leachate is generated, because iron 8-hydroxyquinoline can be removed by strong acids. In addition, 8-hydroxyquinoline is toxic to aquatic organisms and costly (EFSA 2011) and may not be suitable for field application.

Fatty Acids

Fatty acids can make pyrite highly hydrophobic, which may reduce the contact of Fe^{3+} with the surface (Nyavor et al. 1996). Nyavor et al. (1996) studied the effectiveness of fatty acid in a laboratory experiment and found that it significantly inhibited both chemical and biological oxidation. Jiang et al. (2000) used a mono-saturated fatty acid such as sodium oleate to create a hydrophobic layer on pyrite surfaces, but this coating resulted in incomplete oxidation. Application of this chemical on a field scale may not be feasible because it is toxic and costly (Wu et al. 2006).

Oxalic Acid

The oxidation of pyrite is suppressed by coating it with oxalic acid because this treatment lowers the standard redox potential of Fe(III)/Fe(II) and forms a complex (Sasaki et al. 1996). Belzile et al. (1997) studied the possible passivation of pyrite by oxalic acid as compared with other agents such as acetyl acetone, humic acid and sodium silicate on samples from Kidd Creek mine in Timmins, Ontario. They concluded that oxalic acid was more efficient than the other agents tested in reducing oxidation of pre-oxidized samples. However, it was noted that the effectiveness of oxalic acid requires high temperature (65 °C), which restricts its field application; it is also toxic (Belzile et al. 1997).

BDET

The disodium salt of 1, 3-benzenediamidoethanthiol (Na_2BDET) is generally used to bind divalent metals (Matlock et al. 2001). As the thiol-based compounds form iron–sulfur clusters, Osterloh et al. (1998) proposed that Na_2BDET could be utilized to form covalent Fe-BDET linkages along pyrite lattices. Subsequently, Matlock et al. (2003) conducted an experiment on coal refuse using Na_2BDET and found this compound inhibited the dissolution of pyrite. The concentration of Fe was reduced by 99.3% and 97.5% at respective pH values of 6.5 and 3. The authors, hypothesized from this result that BDT ligand was forming a covalent bond with pyritic Fe(II) in the coal. Recently, Zaman et al. (2007) confirmed the effectiveness of BDET for inhibiting sulfide oxidation under varying pH (1, 3 and 7) conditions, over a period of 1 month. These authors hypothesized that the inhibition of oxidation was due to formation of covalent bonds between the BDET ligand and pyritic Fe(II) in the coal. However, this study did not address long-term performance, effects on bacterial oxidation, or environmental impact. In any case, the use of BDET may not be appropriate at the field level because of high reagent costs.

Catechol

Satur et al. (2007a) proposed carrier-microencapsulation (CME) as a technique for suppressing pyrite oxidation. In CME, a water-soluble organic agent is used as a carrier to transport Ti from Ti minerals to pyritic surfaces. The organic carrier then decomposes and forms a thin coating of TiO_2 or $Ti(OH)_4$ on the surface. Satur et al. (2007a, b) studied this technique using catechol (benzene-1, 2-doil) as the carrier, and Ti from TiO_2 as the metal ion. They found that this coating is stable and resistant to both acids and oxidants, and suppressed pyrite oxidation. However, their experiment was conducted at pH 5–6 under a N_2 atmosphere, and was only of short duration (7 days). Therefore, its effectiveness at ambient atmospheric conditions is largely unknown. In nature, a very small quantity of catechol is produced but for the actual application, much higher quantities of catechol are needed, which is very expensive. Another major disadvantage of this technique is that at low pH (<4) the extraction of Ti is limited (Satur et al. 2007a). Jha (2010) studied the CME method using a water-soluble organic carrier-catechol, combined with Si ions (viz., a Si-catechol complex $Si(cat)_3^{2-}$) to control pyrite oxidation in coal and in metal sulfide mineral processing. In this method, pyrite is coated with a thin layer of Si oxide or Si hydroxide, which significantly suppressed pyrite oxidation even in the presence of iron oxidizing bacteria. This technique may be suitable for industrial applications involving mineral processing. However, longer-term testing is needed to determine its performance in commercial applications.

2.4 Electrochemical Cover

ENPAR Technologies, Guelph, Ontario, Canada, (ENPAR Technologies Inc. 2000) patented an electrochemical method that provides a cost effective solution to AMD and can complement conventional strategies (Brousseau et al. 2000). The key to this

Fig. 7 Diagram of an electrochemical protection system (Brousseau et al. 2000)

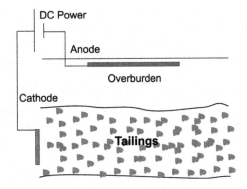

technique involves negatively polarizing the tailing-electrolyte interface so that dissolved O_2 is reduced at the surface of the tailings. In this method, sulfide-bearing rock acts as a cathode in the electrochemical cell. There have been many laboratory and field studies carried out by ENPAR and Shelp et al. (1995), in which they have demonstrated that high sulfide-bearing tailings can act as an electrical conductor, such that the sulfide tailings/deposits act as the cathode of an electrochemical cell as shown in Fig. 7. A field demonstration of this technique was conducted at in the Sudbury basin, Ontario, Canada at a tailings deposit 5 ha in area, and composed primarily of pyrrhotite (~60–70%) (Brousseau et al. 2000; Lin et al. 2001). The results indicated that this technique removed >99.9% of the dissolved oxygen, and after treatment the tailings became thermodynamically more stable. This technology has been effectively applied to a small portion of the Golden Sunlight Mine tailings of Montana, USA (ENPAR Technologies Inc. 2002).

Shelp et al. (2003) studied this technique under laboratory conditions. Their testing was designed to prevent the formation of AMD by reducing the partial pressure of oxygen (PO_2) or dissolved oxygen (DO) available to sulfide-containing mine waste. They reported that, with an input of 45 Amp/ha, the electrochemical cover reduced PO_2 50-fold. However, the mechanism by which this technique works is still unclear, and more field- and lab-based studies will be needed to prove its long-term effectiveness. Whether or not this technique will address the effects of biological oxidation is uncertain.

2.5 Desulfurization

Environmental desulfurization has been used as a potential alternative to manage sulfide-bearing tailings. This process was used to separate sulfide-rich minerals, leaving the majority of low sulfur containing tailings that preferably become non-acid-generating. Desulfurization is based on the principle of froth flotation, which is widely used to separate valuable minerals from waste material. Bois et al. (2004) showed that metal mine tailings, containing about 20% S, could be separated by flotation into materials containing low (0.5%) and high (43%) sulfur. Many researchers have used the desulfurization technique for recovering sulfide minerals from tailings (Broekman and Penman 1991; Leppinen et al. 1997; Humber 1995; Luszczkiewicz and Sztaba 1995; Benzaazoua et al. 1998). Leppinen et al. (1997) successfully used this method to recover sulfide minerals from tailings of the Pyhasalami Cu-Zn mine in Finland. A US Patent (8052774B2) by Nagase et al. (2010) described the desulfurization of copper sulfide minerals for extracting gold. Recently, Benzaazoua et al. (2000) used this method as an alternative to manage Canadian mine tailings. Hesketh et al. (2010) also demonstrated the feasibility of using this technology for desulfurization of cupper sulfide tailings, and achieved a lean content of sulfur in the tailings (0.2% S). Such low-S tailings are classified as being non-acid-producing material.

After desulfurization, low-sulfur tailings (mainly finely fraction) have been used to cover existing high-sulfur tailings (Martin and Fyfe 2011). Such sulfide-rich tailings can be safely processed or managed by using various techniques that reduce subsequent leaching. For example, at a proposed copper mine in Central America, the sulfur-rich tailings are placed under pond water during mining operations (AMEC 2010), and the mine is filled with water at the end of the operation (AMEC 2010). Low-sulfur tailings can be used for other prevention methods, such as soil covers (Sjoberg-Dobchuk et al. 2003; Bussière 2007; MEND 2010). However, when this process is used, the acid generating potential of low sulfur-bearing tailings should be evaluated to ensure that the cover material it is not of an acid-generating type.

3 Summary

AMD is one of the critical environmental problems that causes acidification and metal contamination of surface and ground water bodies when mine materials and/or overburden-containing metal sulfides are exposed to oxidizing conditions. The best option to limit AMD is early avoidance of sulfide oxidation. Several techniques are available to achieve this. In this paper, we review all of the major methods now used to limit sulfide oxidation. These fall into five categories: (1) physical barriers, (2) bacterial inhibition, (3) chemical passivation, (4) electrochemical, and (5) desulfurization. We describe the processes underlying each method by category and then address aspects relating to effectiveness, cost, and environmental impact. This paper may help researchers and environmental engineers to select suitable methods for addressing site-specific AMD problems.

Irrespective of the mechanism by which each method works, all share one common feature, i.e., they delay or prevent oxidation. In addition, all have limitations. Physical barriers such as wet or dry cover have retarded sulfide oxidation in several studies; however, both wet and dry barriers exhibit only short-term effectiveness. Wet cover is suitable at specific sites where complete inundation is established, but this approach requires high maintenance costs. When employing dry cover, plastic liners are expensive and rarely used for large volumes of waste.

Bactericides can suppress oxidation, but are only effective on fresh tailings and short-lived, and do not serve as a permanent solution to AMD. In addition, application of bactericides may be toxic to aquatic organisms.

Encapsulation or passivation of sulfide surfaces (applying organic and/or inorganic coatings) is simple and effective in preventing AMD. Among inorganic coatings, silica is the most promising, stable, acid-resistant and long lasting, as compared to phosphate and other inorganic coatings. Permanganate passivation is also promising because it creates an inert coating on the sulfide surface, but the mechanism by which this method works is still unclear, especially the role of pH. Coatings of Fe-oxyhydroxide, which can be obtained from locally available fly ash are receiving attention because of its low cost, self-healing character, and high cementation capacity. Among organic coatings, lipids and natural compounds such as humic acid appear to be encouraging because they are effective, and have a low environmental impact and cost. Common advantages of organic vs. inorganic coatings are that they work best at low pH and can prevent both chemical and biological oxidation. However, organic coatings are more expensive than inorganic coatings. Furthermore, while organic coatings are effective under laboratory conditions, they often fail under field conditions or require large amounts of reagents to insure effectiveness.

Electrochemical cover technology may become a suitable technique to prevent AMD, but the mechanism by which this technique operates is still under investigation. Limitations of this method include the initial capital cost and ongoing costs of anodes and cathodes.

Desulfurization is an alternative process for managing large-scale sulfide wastes/ tailings. This process can separate sulfide minerals into a low-volume stream, leaving mainly waste with low sulfur content that will be non-acid-generating. The attractiveness of desulfurization is that it is simple and economic.

Our review has clearly disclosed that more information is needed for most of the AMD-mitigation techniques available. Silica passivation has shown promise, but more extensive field-testing is needed to reduce it to commercial viability. Silica is the dominant element in fly ash, and therefore, its use as a low-cost, easily accessible coating should be evaluated. Permanganate passivation also requires further study to understand the role of pH. The secondary formation of Fe-oxyhydroxide minerals from Fe-oxyhydroxides, from the standpoint of their phase transformation, stability and effectiveness, should be assessed over longer experimental periods. All inorganic coatings are designed to inhibit abiotic oxidation of pyrite; however, their effect on biotic pyrite oxidation is not well known and should be further studied. Currently, there is no information available on longer-term field application of organic reagents. Such information is needed to evaluate their lifetime environmental and performance effects. Future studies require spectroscopic analyses of all coating types to achieve a better understanding of their surface chemistry. In addition, a thorough mineralogical and geochemical characterization of waste materials is essential to understand the acid generating potential, which can indeed help to select better prevention measures.

From having performed this review, we have concluded that no single method is technologically mature, although the majority of methods employed are promising for some applications, or at specific sites. Combining techniques can help achieve AMD containment in some cases. For example, applying dry cover (e.g., soil) in combination with liming material or a bactericide, or applying inorganic coatings (e.g., silica) along with organic reagents (e.g., lipids or humic acid) may be more effective than utilizing any single technique alone.

Acknowledgments This work was financially supported by BK21 Advanced Geo-Environment Research Team, Kunsan National University, Korea. We are grateful to the editor and two anonymous reviewers for their valuable comments and suggestions that greatly improved the manuscript.

References

Ačai P, Sorrenti E, Polakovič M, Kongolo M, Donato PD (2009) Pyrite passivation by humic acid investigated by inverse liquid chromatography. Colloid Surface Physicochem Eng Aspect 337:39–46

Akabzaa TM, Armah TEK, Baneong-Yakubo BK (2007) Prediction of acid mine drainage generation potential in selected mines in the Ashanti Metallogenic Belt using static geochemical methods. Environ Geol 52:957–964

Akcil A, Koldas S (2006) Acid mine drainage (AMD): causes, treatment and case studies. J Cleaner Product 14:1139–1145

Al TA, Martin CJ, Blowes DW (2000) Carbonate–mineral/water interaction in sulfide-rich mine tailings. Geochim Cosmochim Acta 64:3933–3948

AMEC (2010) Mina de cobre Panamá project feed study — executive summary, minera panama, S.A. (http://www.inmetmining.com/Theme/Inmet/files/Section%200Executive%20Summary_FinalFEED_31%20March.pdf)

Beck SW (2003) Passivation of weathered and fresh sulfidic rock, A thesis for the degree of Master of Science in Metalurrical Engineering, Unv of Neneda, Reno

Bacelar-Nicolau P, Johnson DB (1999) Leaching of pyrite by acidophilic heterotrophic iron-oxidizing bacteria in pure and mixed cultures. Appl Environ Microbiol 65(2):585–590

Backes CA, Pulford ID, Duncan HJ (1987) Studies on the oxidation of pyrite in colliery spoil: inhibition of the oxidation by amendment treatments. Recl Reveg Res 6:1–11

Bell FG, Bullock SET, Hälbich TFJ, Londsay P (2001) Environmental impacts associated with an abandoned mine in the Witbank Coalfield, South Africa. Int J Coal Geol 45:195–216

Belzile N, Maki S, Chen YW, Goldsack D (1997) Inhibition of pyrite oxidation by surface treatment. Sci Total Environ 196:177–186

Benzaazoua M, Bussière B, Kongolo M, McLaughlin J, Marion P (2000) Environmental desulphurization of four canadian mine tailings using froth flotation. Int J Mineral Processing, 60:57–74

Bessho M, Wajima T, Ida T, Nishiyama T (2011) Experimental study on prevention of acid mine drainage by silica coating of pyrite waste rocks with amorphous silica solution. Environ Earth Sci 64:311–318

Benzaazoua M, Bussière B, Lelièvre J (1998) Flotation non-sèlective des minèraux sulphurès appliqué dans la gestion environmntale des rejects miniers. Proceedings of Canadian Minetal Processors, Ottawa (Canada):682–695

Black A, Craw D (2001) Arsenic, copper and zinc occurrence at the Wangaloa coal mine, southeast Otago, New Zealand. Int J Coal Geol 45:181–193

Blowes DW, Reardon EJ, Jambor JL, Cherry JA (1991) The formation and potential importance of cemented layers in inactive sulphide mine railings. Geochim Cosmochim Acta 55:965–978

Bois D, Poirier P, Benzaazoua M, Bussière BA (2004) Feasibility study on the use of desulphurized tailings to control acid mine drainage. In: Proceedings 2004—36th annual meeting of the Canadian mineral processors, January 20–22, 2004, Ottawa, ON, pp 361–380

Brady KBC, Smith MW, Beam RL, Cravotta CA (1990) Effectiveness of the addition of alkaline materials at surface coal mines in preventing or abating acid mine drainage—part 2. Mine site case studies. In: Skousen J, Sencindiver J, Samuel D (eds) Proceedings of the 1990 mining and reclamation conference and exhibition, Charleston, West Virginia, April 23–26, 1990, Morgantown, W.Va, West Virginia University Publication, vol 1, pp 227–241

Broekman BR, Penman DW (1991) The Prieske experience: flotation development in copper-zinc separation. J S Afr Inst Min Metall 91:257–265

Brousseau JHR, Seed LP, Lin MY, Shelp GS, Fyfe JD (2000). In: Singhal RK, Mehrotra AK (eds) Proceedings sixth international conference on environmental issues and management of waste in energy and mineral production: SWEMP 2000, Calgary, Alberta, Canada

Brown M, Barley B, Wood H (2002) Minewater treatment: technology, application and policy. International Water Association Publishing, London

Bulusu S, Aydilek AH, Rustagi N (2007) CCB-based encapsulation of pyrite for remediation of acid mine drainage. J Hazard Mater 143:609–619

Bussière B (2007) Colloquium 2004: hydrogeotechnical properties of hard rock tailings from metal mines and emerging geoenvironmental disposal practices. Can Geotech J 44:1019–1052

Bussiere B, Benzaazoua M, Aubertic M, Mbonimpa M (2004) A laboratory study of covers made of low-sulfide tailings to prevent acid mine drainage. Environ Geol 45:609–622

Cai MF, Dang Z, Chen YW, Belzile N (2005) The passivation of pyrrhotite by surface coating. Chemosphere 61:659–667

Caldeira CL, Ciminelli VST, Asare KO (2010) The role of carbonate ions in pyrite oxidation in aqueous systems. Geochim Cosmochim Acta 74:1777–1789

Campbell RN, Lindsay P, Clemens AH (2001) Acid generating potential of waste rock and coal ash in New Zealand coal mines. Int J Coal Geol 45:163–179

Cárdenes V, Eynde VD, Paradelo R, Monterroso C (2009) Passivation techniques to prevent corrosion of iron sulfides in roofing slates. Corrosion Sci 51:2387–2392

Caruccio FT (1983) The effect of plastic liner on acid loads: DLM site. In: Proceedings, fourth annual West Virginia surface mine drainage task force symposium, Morgantown, WV, 26 May 1983

Caruccio FT, Ferm JC (1974) Paleoenvironment—predictor of acid mine drainage problems. In: Proceedings of the 5th coal mine drainage research symposium, National Coal Association (USA), Kentucky, pp 5–9

Casiot C, Leblanc M, Bruneel O, Personne JC, Koffi K, Elbaz-poulichet F (2003a) Formation of As-rich waters within a tailings impoundment (Carnoulès, France). Aquatic Geochem 9:273–290

Casiot C, Morin G, Juillot F, Bruneel O, Personne J, Leblanc M, Duquesne K, Bonnefoy V, Elbaz-Poulichet F (2003b) Bacterial immobilization and oxidation of arsenic in acid mine drainage (Carnoulès creek, France). Water Res 37:2929–2936

Chen YW, Yuerong L, Cai MF, Belzile N, Dang Z (2006) Preventing oxidation of iron sulfide minerals by polyethylene polyamines. Mineral Eng 19:19–27

Chon H, Hwang J (2000) Geochemical characteristics of the acid mine drainage in the water system in the vicinity of the Dogye coal mine in Korea. Environ Geochem Health 22:155–172

Cravotta CA, Brightbill RA, Langland MJ (2010) Abandoned mine drainage in the swatara creek basin, Southern Anthracite Coalfield, Pennsylvania, USA: 1. Stream water quality trends coinciding with the return of fish, USGS Published Report, University of Nebraska—Lincoln

Dave NK, Vivyurka AJ (1994) Water cover on acid generating uranium tailings—laboratory and field studies. In: Proceedings of 4th international conference on acid rock drainage, vol 1, pp 297–306

Davé NK, Lim TP, Horne D, Boucher Y, Stuparyk R (1997) Water cover on reactive tailings and waste rock: laboratory studies of oxidation and metal release characteristics. In: Proceedings of the fourth ICARD, Vancouver, pp 779–794

De Vries NHC (1996) Process for treating iron-containing sulfide rocks and ores. US Patent 5,587,001

Demers I, Bussière B, Benzaazoua M, Mbonimpa M, Blier A (2008) Column test investigation on the performance of monolayer covers made of desulphurized tailings to prevent acid mine drainage. Mineral Eng 21:317–329

Dugan PR (1987) Prevention of formation of acid drainage from high-sulfur coal refuse by inhibition of iron and sulfur-oxidizing microorganisms: II. Inhibition, in "Run of Mine" refuse under simulated field condition. Biotechnol Bioeng 29:29–54

Dugan PR, Apel WA (1983) Bacteria and acidic drainage from coal refuse: inhibition by sodium lauryl sulfate and sodium benzoate. Appl Environ Microbiol 46:279

Duval JFL, Sorrenti E, Waldvogel Y, Gorner T, Donato PD (2007) On the use of electrokinetic phenomena of the second kind for probing electrode kinetic properties of modified electron-conducting surfaces. Phys Chem Chem Phys 9:1713–1729

Eger P, Antonson D (2002) "Use of microencapsulation to prevent acid rock drainage", report to MSE technology applications. Minnesota Department of Natural Resources, St. Paul, Minnesota

Eger P, Antonson D (2004) Use of microencapsulation to prevent acid rock drainage. Minnesota Department of Natural Resources, St. Paul, Minnesota

Eger P, Mitchell P (2007) "The use of microencapsulation to prevent acid rock drainage," presented at mining and the environment IV conference, Sudbury, Ontario, Canada, 19–27 Oct 2007

Egiebor NO, Oni B (2007) Acid rock drainage formation and treatment: a review. Asia Pacific J Chem Eng 2:47–62

Elsetinow A (2006) Method for inhibiting oxidation of metal sulfide-containing material. Patent 7,153,541

Elsetinow AR, Schoonew AA, Strongin DR (2001) Aqueous geochemical and surface science investigation on the effect on phosphate on pyrite oxidation. Environ Sci Technol 35:2252–2257

Elsetinow AR, Borda MJ, Schoonen MAA, Strongin DR (2003) Suppression of pyrite oxidation in acidic aqueous environments using lipids having two hydrophobic tails. Adv Environ Res 7:969–974

ENPAR Technologies Inc. (2000) Electrochemical cover for mine wastes. Patent pending, WO 01/38233 A1

ENPAR Technologies Inc. (2002) Innovative amdel electrochemical cover technology proceeds to field testing at the golden sunlight mine. USEPA, Montana, USA

Equeenuddin SM, Tripathy S, Sahoo PK, Panigrahi MK (2010) Hydrogeochemical characteristics of acid mine drainage and water pollution. J Geochem Explor 105:75–82

Equeenuddin SM, Tripathy S, Sahoo PK, Panigrahi MK (2013) Metal behavior in sediments associated with acid mine drainage stream: role of pH. J Geochem Explor 124:230–237

European Food Safety Authority (EFSA), Parma, Italy (2011) Conclusion on the peer review of the pesticide risk assessment of the active substance 8-hydroxyquinoline. EFSA J 9:1964

Evangelou VP (1994) Microencapsulation of pyrite by artificial inducement of $FePO_4$ coatings. In: Proceedings of second international conference on the abatement of acid drainage, Pittsburgh, PA, 24–29 April. United States Bureau of Mines Special Publication SP 06A-94, 2, 321

Evangelou VP (1995a) Pyrite oxidation and its control. CRC, New York

Evangelou VP (1995b) Potential microencapsulation of pyrite by artificial inducement of ferric phosphate coatings. J Environ Qual 24:535–542

Evangelou VP (1996) Oxidation proof silica surface coating iron sulfides. US Patent 5,494,703

Evangelou VP (2001) Pyrite microencapsulation technologies: principles and potential field application. Ecol Eng 17:165–178

Evangelou VP, Zhang YL (1995) A review: pyrite oxidation mechanisms and acid mine drainage prevention. Crit Rev Environ Sci Technol 25:141–199

Felipe V (2008) Permanganate passivation: a study of the longevity of the process and its behavior under different external conditions. M.S. thesis, University of Nevada, Reno, 72 pages; 1460786

Fernandes HM, Franklin MR (2001)Assessment of acid rock drainage pollutants release in the uranium mining site of Poços de Caldas – Brazil. J Environ Radioact 54:5–25

Filion MP, Sirois LL, Ferguson K (1990) Acid mine drainage research in Canada. CIM Bull 83:33–40

Fugill RJ, Sencindiver JC (1986) Effect of topsoil and vegetation on the generation of acid mine drainage from coal refuse. In: Proceedings seventh annual WV surface mine drainage task force symposium, West Virginia mining and reclamation association, Charleston, WV

Gazea B, Adam K, Kontopoulos A (1996) A review of passive system for the treatment of acid mine drainage. Mineral Eng 9:23–42

Geidel G, Caruccio FT (1984) A field evaluation of the selective placement of acidic material within the backfill of a reclaimed coal mine. In: Proceedings of the 1984 symposium on surface mining, sedimentology and reclamation, University of Kentucky, Lexington, KY, pp 127–131

Geller W, Klapper H, Schultze M (1998) Natural and anthropogenic sulfuric acidification of lakes. In: Klapper W, Salomons W, Geller W (eds) Acidic mining lakes. Springer, Berlin, pp 3–15

Georgopoulou ZJ, Fytas K, Soto H, Evangelou B (1996) Feasibility and cost of creating an iron-phosphate coating on pyrrhotite to prevent oxidation. Environ Geol 28:61–69

Glover R (2007) Permanganate passivation of pyrite containing ores: scale up and characterization. MS thesis. University of Nevada, Reno; UMI number 1446427. http://www.docin.com/p-232211543.html

Gomes MEP, Favas PJC (2006) Mineralogical controls on mine drainage of the abandoned Ervedosa tin mine in north-eastern Portugal. Appl Geochem 21:1322–1334

Goel PK (2006) Water pollution causes effects and control, 2nd edn. New Age International Publishers, New Delhi

Groudev S, Georgiev P, Spasova I, Nicolova M (2008) Bioremediation of acid mine drainage in a uranium deposit. Hydrometallurgy 94:93–99

Grout JA, Levings CD (2001) Effects of acid mine drainage from an abandoned copper mine, Britannia Mines, Howe Sound, British Columbia, Canada, on transplanted blue mussels (*Mytilus edulis*). Marine Environ Res 51:265–288

Güler T (2005) Dithiophosphinate–pyrite interaction: voltametry and DRIFT spectroscopy investigations at oxidizing potentials. J Colloid Interface Sci 288:319–324

Hallberg RO, Granhagen JR, Liljemark A (2005) A fly ash/biosludge dry cover for the mitigation of AMD at the falun mine. Chemie der Erde 65:43–63

Hao J, Cleveland C, Lim E, Strongin DR, Schoonen MAA (2006) The effect of adsorbed lipid on pyrite oxidation under biotic conditions. Geochem Trans 7:8

Hao J, Murphy R, Lim E, Schoonen MAA, Strongin DR (2009) Effects of phospholipid on pyrite oxidation in the presence of autotrophic and heterotrophic bacteria. Geochim Cosmochim Acta 73:4111–4123

Harries JR, Ritchie AIM (1985) Pore gas composition in waste-rock dumps undergoing pyrite oxidation. Soil Sci 140:143–152

Harries JR, Ritchie AIM (1987) The effect of rehabilitation on the rate of oxidation of pyrite in a mine waste-rock dump. Environ Geochem Health 9:27–36

Harrington JG (2001) US20016196765

Harris DL, Lottermoser BG (2006a) Evaluation of phosphate fertilizers for ameliorating acid mine waste. Appl Geochem 21:1216–1225

Harris DL, Lottermoser BG (2006b) Phosphate stabilization of polyminerallic mine wastes. Mineralog Mag 70:1–13

Harris DL, Lottermoser BG, Duchesne J (2003) Ephemeral acid mine drainage at the Montalbion silver mine, north Queensland. Aust J Earth Sci 50:797–809

Hesketh AH, Broadhurst JL, Harrison STL (2010) Mitigating the generation of acid mine drainage from copper sulfide tailings impoundments in perpetuity: a case study for an integrated management strategy. Mineral Eng 23:225–229

Hodges G, Roberts DW, Marshall SJ, Dearden JC (2006) The aquatic toxicity of anionic surfactants to Dophnia magna—a comparative QSAR study of linear alkylbenzene sulphonates and ester sulphonates. Chemosphere 63:1443–1450

Holmström H, Salmon UJ, Carlsson E, Petrov P, Öhlzander B (2001) Geochemical investigations of sulfide bearing tailings at Kristineberg, northern Sweden, a few years after remediation. Sci Total Environ 273:111–133

Huang X, Evangelou VP (1994) Suppression of pyrite oxidation rate by phosphate addition. In: Alper CN, Blowes DW (eds) Environmental geochemistry of sulfide oxidation. American Chemical Society, Washington, DC, pp 562–573

Huminicki DMC, Rimstidt JD (2009) Iron oxyhydroxide coating of pyrite for acid mine drainage control. Appl Geochem 24:1626–1634

Humber AJ (1995) Separation of sulphide minerals from mill tailings, Sudbury 95, Conference on mining and the environment, Sudbury, Ontario, pp. 149–158

Hurst S, Schneider P, Meinrath G (2002) Remediating 700 years of mining in saxony: a heritage from ore mining. Mine Water Environ 21:3–6

Ingledew WJ (1982) Thiobacillus ferrooxidans the bioenergetics of an acidophilic chemolithotroph. Biochim Biophys Acta 683:89–117

Jennings SR, Dollhopf DJ, Inskeep WP (2000) Acid production from sulfide minerals using hydrogen peroxide weathering. Appl Geochem 15:235–243

Jha RKT (2010) Carrier micro-encapsulation using Si and catechol to suppress pyrite flotation and oxidation. Ph.D. thesis, Hokkaido University, Sapporo, Japan

Ji SW, Cheong YW, Yim GJ, Bhattacharya J (2007) ARD generation and corrosion potential of exposed roadside rockmass at Boeun and Mujoo, South Korea. Environ Geol 52:1033–1043

Ji MK, Gee ED, Yun HS, Lee WR, Park YT, Khan MA, Jeon BH, Choi J (2012) Inhibition of sulfide mineral oxidation by surface coating agents: Batch and field studies J Hazard Material 229–230: 298–306

Jiang CL, Wang XH, Parekh BK (2000) Effect of sodium oleate on inhibiting pyrite oxidation. Int J Miner Process 58:305–318

Johnson DB, Hallberg KB (2005) Acid mine drainage remediation options: a review. Sci Total Environ 338:3–14

Kapadia PC, Arnoid J, Marshal GP, Thompson JS (1999) Preventing acid rock drainage using Dupont's surface passivation technology. In: Kosich D, Miller G (eds) Closure, remediation and management of precious metals heap leach facilities

Kargbo D, Chatterjee S (2005) Stability of silicate coatings on pyrite surfaces in a low pH environment. J Environ Eng 131:1340–1349

Kargbo DM, Fanning DS, Inyang HI, Duell RW (1993) The environmental significance of acid sulfate clays as waste covers. Environ Geol 22:218–226

Kargbo DM, Atallah G, Chatterjee S (2004) Inhibition of pyrite oxidation by a phospholipids in the presence of silicate. Environ Sci Technol 38:3432–3441

Khummalai N, Boonamnuayvitaya V (2005) Suppression of arsenopyrite surface oxidation by sol-gel coatings. J Biosci Bioeng 99:277–284

Kirk-Othmer (1984) Encyclopedia of Chem Technol, vol 24, 3rd edn, pp 645–661

Kleinmann RLP (1989) Acid mine drainage in the United States: controlling the impact on streams and rivers. In: 4th world congress on the conservation of builts and natural environments. University of Toronto. pp 1–10

Kleinmann RLP (1990) At-source of acid mine drainage. Mine Water Environ 9:85–96

Kleinmann RLP (1998) Bactericidal control of acidic drainage. In: Brady KC, Smith MW, Schueck J (eds) Coal mine drainage prediction and pollution prevention in Pennsylvania, PA DEP, Harrisburg, PA, 15:1–6

Kleinmann RLP (1999) Bactericidal control of acidic drainage. In: Coal mine drainage prediction and pollution prevention in Pennsylvania. The Pennsylvania Department of Environmental Protection, Chapter 15, pp 15-1 to 15-6 (http://www.dep.state.pa.us/dep/deputate/minres/districts/cmdp/main.htm)

Kleinmann RLP, Crerar DA (1979) Thiobacillus ferroxidans and the formation of acidity in simulated coal mine environments. Geomicrobiol J 1:373–388

Kleinmann RLP, Erickson PM (1981) Field evaluation of a bactericidal treatment to control acid drainage. In: Graves DH (ed) Proceedings of the symposium on surface mining hydrology, sedimentology and reclamation. University of Kentucky, Lexington, KY, pp 325–329

Kleinmann RLP, Erickson PM (1983) Control of acid mine drainage from coal refuse using anionic surfactants. Report of investigation No. 8847, U.S Bureau of Mines

Kleinmann RLP, Crerar DA, Pacelli RP (1981) Biogeochemistry of acid mine drainage and a method to control acid formation. Mining Eng 33:300–305

Lalvani SB, Deneve BA, Weston A (1990) Passivation of pyrite due to surface treatment. Fuel 69:1567–1569

Lalvani SB, Deneve BA, Weston A (1991) Prevention of pyrite dissolution in acid media. Corrosion 47:55–61

Lalvani SB, Zhang G, Lalvani LS (1996) Coal pyrite passivation due to humic acids and lignite treatment. Fuel Sci Technol Int 14:1291–1313

Lan Y, Huang X, Deng B (2002) Suppression of pyrite oxidation by iron 8-hydroxyquinoline. Arch Environ Contam Toxicol 43:168–174

Langworthy TA (1978) Microbial life in extreme pH values. In: Kuschner DJ (ed) Microbial life in extreme environments. Academic, New York, pp 279–315

Lapakko KA (1994) Subaqueous disposal of mine waste: laboratory investigation. Bureau of Mines Special Publication, SP 06 A-94, pp 270–278

Lattuada RM, Menezes CTB, Pavei PT, Peralba MCR, Dos Santos JHZ (2009) Determination of metals by total reflection X-ray fluorescence and evaluation of toxicity of a river impacted by coalmining in the south of Brazil. J Hazard Mater 163:531–537

Lei L, Song C, Xie X, Wang F (2010) Acid mine drainage and heavy metal contamination in groundwater of metal sulfide mine at arid territory (BS mine, Western Australia). Trans Non Ferrous Met Soc China 20:1488–1493

Leppinen JO, Salonsaari P, Palosaari V (1997) Flotation in acid mine drainage control: beneficiation of concentrate. Canadian Metallurgical Quarterly 36:225–230

Lesage S, Xu H, Novakowshi KS, Brown S, Durham L (1995) Use of humic acids to enhance the removal of aromatic hydrocarbons from contaminated aquifers. Part II: pilot scale. In: Proceedings of the 5th annual symposium on groundwater and soil remediation. Toronto, ON, 2–6 Oct 1995

Lin M, Seed L, Yetman D, Fyfe J, Chesworth W, Shelp G (2001) Electrochemical cover technology to prevent the formation of acid mine drainage. In: Proceedings of the 25th annual British Columbia mine reclamation symposium in Campbell River, BC

Lindgren P, Parnell J, Holm GN, Broman C (2011) A demonstration of an affinity between pyrite and organic matter in a hydrothermal setting. Geochemical Transact 12:3

Liwarska BE, Miksch K, Malachowska JA, Kalka J (2005) Acute toxicity and genotoxicity of five selected anionic and nonionic surfactants. Chemosphere 58:1249–1253

Lollar BS (2005) Environmental geochemistry. Elsevier Publisher, Oxford

Loos MA, Bosch C, Mare J, Immelman E, Sanderson RD (1989) Evaluation of sodium lauryl sulfate, sodium benzoate and sorbic acid as inhibitors of acidification of South African coal waste. In: Groundwater and mining: proceedings of the 5th biennial symposium of the groundwater division of the geological survey of South Africa Randburg, Transvaal, Pretoria, Geological Society of South Africa, pp 193–200

Lottermoser B (2007) Mine wastes characterization, treatment and environmental impacts, 2nd edn. Springer Publisher, Heidelberg

Lusardi PJ, Erickson PM (1985) Assessment and reclamation of an abandoned acid-producing strip mine in northern Clarion County, Pennsylvania. In: Proceedings of the 1985 symposium surface mining hydrology, sedimentology, and reclamation, University of Kentucky, Lexington, KY, pp 313–321

Luszckiewicz A, Sztaba KS (1995) Beneficiation of flotation tailing from polish copper sulfide ores 25(4):121–124

Mallo SJ (2011) The menace of acid mine drainage: an impending challenge in the mining of Lafia-obi coal, Nigeria. Continental J Eng Sci 6:46–54

Martin J, Fyfe J (2011) Innovative closure concepts for xstrata nickel onaping operations, presentation in Sudbury 2011, mining and the environment international conference V, 25–30 June 2011

Matlock MM, Howerton BS, Henkee KR, Atwood DA (2001) Irreversible binding of mercury and lead from aqueous systems with a newly designed multi dentate ligand. J Hazard Mat B 84:73–82

Matlock MM, Howerton BS, Atwood DA (2003) Covalent coating of coal refuses to inhibit leaching. Adv Environ Res 7:495–501

Mauric A, Lottermoser BG (2011) Phosphate amendment of metalliferous waste rocks, Century Pb-Zn mine, Australia: laboratory and field trials. Appl Geochem 26:45–56

McCloskey AL (2005) Prevention of acid mine drainage generation from open-pit highwalls—final report. Mine waste technology program activity III, project 26, EPA/600/R-05/060

Miller G, Van Zyl D (2008) Personal communication

Miller S, Smart R, Andrina J, Neale A, Richards D (2003) Evaluation of limestone covers and blends for long-term acid rock drainage control at the grasberg mine, papua province, Indonesia. In: Proceedings of 6th international conference on acid rock drainage (ICARD), July 12–18, Cairns, QLD, Australia, AusIMM, pp 133–141

Miller S, Rusdinar Y, Smart R, Andrina J, Richards D (2006) Design and construction of limestone blended waste rock dumps—lessons learned from a 10-year study at Grasberg. In: Barnhisel RI (ed) Proceedings of 7th international conference on acid rock drainage (ICARD), St. Louis, MO, American Society of Mining and Reclamation, Lexington, KY, 26–30 Mar 2006

Mine Environment Neutral Drainage Program (MEND) (2001) Prevention and control, vol 4. manual 5.4.2d. In: Tremblay GA, Hogan CM (eds) CANMET

Mine Environment Neutral Drainage Program (MEND) (2010) Evaluation of the water quality benefits from encapsulation of acid-generating tailings by acid-consuming tailings—December

Montero IC, Brimhall GH, Alpers CN, Swayze GA (2005) Characterization of waste rock associated with acid drainage at the Penn Mine, California, by ground-based visible to short-wave infrared reflectance spectroscopy assisted by digital mapping. Chem Geol 215:452–472

Mosquera MJ, Pozo J, Esquivias L, Rivas T, Silva B (2002) Application of mercury porosimetry to the study of xerogels used at stone consolidants. J Non-Crystal Solids 311:185–194

Mylona E, Xenidis A, Paspaliaris I (2000) Inhibition of acid generation from sulfidic wastes by the addition of small amounts of limestone. Mineral Eng 13:1161–1175

Naicker K, Cukrowska E, McCarthy TS (2003) Acid mine drainage arising from gold mining activity in Johannesburg, South Africa and environs. Environ Pollut 122:29–40

Nagase N, Asano S, Takano M, Takeda K, Heguri S, Idegami A (2010) Method for concentration of gold in copper sulfide minerals A US Patent No US8052774

Nganje TN, Adamu CI, Ntekim EEU, Ugbaja AN, Neji P, Nfor EN (2010) Influence of mine drainage on water quality along River Nyaba in Enugu South-Eastern Nigeria. Afr J Environ Sci Technol 4:132–144

Nicholson RV, Gillham RW, Reardon EJ (1988) Pyrite oxidation in carbonate– buffered solutions: 1. Experimental kinetics. Geochim Cosmochim Acta 52:1077–1085

Nicholson RV, Gillham RW, Cherry JA, Reardon EJ (1989) Reduction of acid generation in mine tailings through use of moisture-retaining cover layers as oxygen barriers. Can Geotech J 26:1–8

Nicholson RV, Gillham RW, Reardon EJ (1990) Pyrite oxidation in carbonate buffered solutions: 2. Rate control by oxide coatings. Geochim Cosmochim Acta 54:395–402

Nieto JM, Sarmiento AM, Olías M, Canovas CR, Riba I, Kalman J, Delvalls TA (2007) Acid mine drainage pollution in the Tinto and Odiel rivers (Iberian Pyrite Belt, SW Spain) and bioavailability of the transported metals to the Huelva Estuary. Environ Int 33:445–455

Nordstrom KD (1982) Aqueous pyrite oxidation and the consequent formation of secondary minerals. In: Kittrick JA, Fanning DS, Hossner LR (eds) Acid sulfate weathering pedogeochemistry and relationship to manipulation of soil materials. Soil Science Society, America Press, Madison, WI, pp 37–56

Nordstrom DK, Alpers CN (1999) Negative pH, efflorescent mineralogy, and consequences for environmental restoration at the Iron Mountain Superfund site, California. Proc Natl Acad Sci U S A 96:3455–3462

Nordstrom DK, Alpers CN, Ptacek CJ, Blowes DW (2000) Negative pH and extremely acidic mine waters from Iron Mountain, California. Environ Sci Technol 34:254–258

Nyavor K, Egiebor NO (1995) Control of pyrite oxidation by phosphate coating. Sci Total Environ 162:225–237

Nyavor K, Egiebor NO, Fedorak PM (1996) Suppression of mineral pyrite oxidation by fatty acid amine treatment. Sci Total Environ 182:75–83

Olson GJ, Clark TR, Mudder TI (2005) Acid rock drainage prevention and treatment with thiocyanate and phosphate containing materials. National Meetings of the American Society of Mining and Reclamation, Breckenridge, CO, June 19–23, 2005. Published by ASMR, Lexington, KY, pp 831–841

Onysko SJ, Kleinmann RLP, Erickson PM (1984) Ferrous iron oxidation by Thiobacillus ferroxodans: inhibition with benzoic acid, sorbic acid, and sodium lauryl sulfate with benzoic acid, sorbic acid, and sodium lauryl sulfate. Appl Environ Microbiol 48:229–231

Osterloh F, Saak W, Pohl S, Kroeckel M, Meier C, Trautwein AX (1998) Synthesis and characterization of neutral hexanuclear iron sulfur clusters containing stair-like [Fe6(u3-S)4(u2-SR)4] and nest-like [Fe6(u3-S)2(u2-S)2(u4-S)(u2-SR)4]. Inorg Chem 37:3581–3587

Palamo A, Fernaddez-Jimenez A, Kovalchuk G, Ordonez LM, Naranjo MC (2007) Opc-fly ash cementitious systems: study of gel binder produced during alkaline hydration. J Mater Sci 42:2958–2966

Pandey S, Yacob TW, Silverstein J, Rajaram H, Minchow K, Basta J (2011) Prevention of acid mine drainage through complexation of ferric iron by soluble microbial growth products. American Geophysical Union, Fall Meeting 2011, abstract #H43J-1370

Parisi D, Horneman J, Rastogi V (1994) Use of bactericides to control acid mine drainage from surface operations. In: Proceedings of the international land reclamation and mine drainage conference, U.S. Bureau of Mines SP 06B-94, Pittsburgh, PA, pp 319–325

Pedersen TF, Mueller B, Pelletier CA (1991) On the reactivity of submerged mine tailings in fjord and lake in British Columbia. In: Gadsby JW, Malick JA (eds) Acid mine drainage: designing for closure. Bi Tech Publishers, Vancouver, BC, pp 281–293

Pedersen TF, Mueller B, McNee JJ, Pelletier CA (1993) The early diagenesis of submerged sulfide-rich mine tailings in Anderson lake, Manitoba. Can J Earth Sci 30:1099–1109

Pelo DS, Musu E, Cidu R, Frau F, Lattanzi P (2009) Release of toxic elements from rocks and mine wastes at the Furtei gold mine (Sardnia, Italy). J Geochem Explor 100:142–152

Peppas A, Komnitsas K, Halikia I (2000) Use of organic covers for acid mine drainage control. Mineral Eng 13:563–574

Pérez-López R, Jordi C, Miguel NJ, Carlos A (2005) The role of iron coating on the oxidative dissolution of a pyrite-rich sludge. In: Loredo J, Pendas F (eds) Ninth international mine water congress, pp 575–579

Pérez-López R, Cama J, Nieto JM, Ayora C (2007a) The iron-coating role on the oxidation kinetics of a pyritic sludge doped with fly ash. Geochim Cosmochim Acta 71:1921–1934

Pérez-López R, Nieto JM, Almodovar GR (2007b) Utilization of fly ash to improve the quality of the acid mine drainage generated by oxidation of a sulfide-rich mining waste: column experiments. Chemosphere 67:1637–1646

Pérez-López R, Nieto JM, Alvared-valero AM, Almodovar GR (2007c) Mineralogy of the hardpan formation processes in the interface between sulfide-rich sludge and fly ash: applications for acid mine drianage mitigation. Am Mineral 92:1966–1977

Pond AP, White SA, Milczarek M, Thompson TL (2005) Accelerated weathering of biosolid-amended copper mine tailings. J Environ Qual 34:1293–1301

Rampe JJ, Runnells DD (1989) Contamination of water and sediment in a desert stream by metal from an abandoned gold mine and mill, Eureka District, Arizona, USA. Appl Geochem 4:445–454

Rao AV, Kulkarni MM, Amalnerkar DP, Seth T (2003) Superhydrophobic silica aerogels based on methyltrimethoxysilane precursor. J Non-Crystal Solids 330:187–195

Reardon EJ, Poscente PJ (1984) A study of gas compositions in sawmill waste deposits: an evaluation of the use of wood waste in close-out of pyrite tailings. Reclam Revegetation Res 3:109–128

Ribet I, Ptacek CJ, Blowes DW, Jambor JL (1995) The potential for metal release by reductive dissolution of weathered mine tailings. J Contamin Hydrol 17:239–273

Rich DH, Hutchison KR (1994) Coal refuse disposal using engineering design and lime chemistry. In: International land reclamation and mine drainage conference, 24–29 April 1994, USDI, Bureau of Mines SP 06A-94, Pittsburgh, PA, pp 392–399

Ritsema CJ, Groenenberg JE (1993) Pyrite oxidation, carbonate weathering, and gypsum formation in a drained potential acid sulfate soil. Soil Sci Soc Am J 57:968–976

Sahoo PK, Tripathy S, Equeendduin SM, Panigrahi MK (2012) Geochemical characteristics of coal mine discharge vis-à-vis behavior of rare earth elements at Jaintia hills coalfield, northeastern India. J Geochem Explor 112:235–243

Sáinz A, Grande JA, De la Torre ML (2004) Characterization of heavy metal discharge into the Ria of Huelva. Environ Int 30:557–566

Sand W, Jozsa PG, Kovacs ZM, Săsăran N, Schippers A (2007) Long-term evaluation of acid rock drainage mitigation measures in large lysimeters. J Geochem Explor 92:205–211

Sasaki K, Tsunekawa M, Tanaka S, Konno H (1996) Supression of microbially mediated dissolution of pyrite by originally isolated fulvic acids and related compounds. Colloid Surface Physicochem Eng Aspect 119:241–253

Satur J, Hiroyoshi N, Ito M, Tsunekawa M (2007b) Carrier-microencapsulation for suppressing floatability and oxidation of pyrite in copper mineral processing. In: Proceedings of the COM2007 — 46th conference of metallurgists hosting Cu 2007 the sixth international copper-cobre conference, Toronto, Canada, Mineral Processing, vol 2, pp 25–30

Satur J, Hiroyoshi N, Tsunekawa M, Mayumi I, Okamoto H (2007b) Carrier–microencapsulation for preventing pyrite oxidation. Int J Mineral Process 83:116–124

Shang JQ, Wang HL, Kovac V, Fyfe J (2006) Site-specific study stabilization of acid generating mine tailing using coal fly ash. J Mater Civil Eng 18:140–150

Sharma P (2010) Acid mine drainage (AMD) and its control. Lambert Academic Publishing, Germany

Sheibach RB, Williams RE, Genes BR (1982) Controlling acid mine drainage from the Picher mining district Oklahoma, US. Int J Mine Water 1:45–52

Shelp G, Chesworth W, Spiers G, Liu L (1995) Cathodic protection of a weathering orebody. In: Hynes TP, Blanchette MC (eds) Proceedings of Sudbury '95 — mining and the environment, May 28–June 1, 1995, Sudbury, Ontario, Canada, vol 3, pp 1035–1042

Shelph ML, Hyward GL, Seed LP, Shelp GS (2003) Electrochemical cover for the prevention of acid mine drainage: a laboratory test. In: Proceedings of the Sudbury 2003 mining and the environment conference, 25–28 May 2003, Sudbury, ON, Canada. Laurentian University, Sudbury, ON, Canada

Silver M, Ritcey M (1985) Effects of iron-oxidizing bacteria and vegetation on acid generation in laboratory lysimeter tests on pyrite-containing uranium tailings. Hydrometallurgy 15:255–264

Simms PH, Yanful EK, St-Arnaud L, Aube B (2000) A laboratory evaluation of metal release and transport in flooded pre-oxidized mine tailings. Appl Geochem 15:1245–1263

Singer PC, Stumm W (1970) Acidic mine drainage — the rate-determining step. Science 167:1121–1123

Sjoberg-Dobchuk B, Wilson GW, Aubertin M (2003) Evaluation of a single-layer desulfurised tailings cover. In: Proceedings of 6th international conference acid rock drainage (ICARD), Cairns, QLD, Australia, AusIMM, 14–17 July 2003

Skousen J, Foreman J (2000) Water management techniques for acid mine drainage control. Green Lands 30(Winter)

Sposito G (1984) The surface chemistry of soils. Oxford University Press, New York

Swanson DA, Barbour SL, Wilson GW (1997) Dry-site versus wet-site cover design. In: Proceedings of the 4th international conference on acid rock drainage, May 30–June 6, Vancouver, BC, vol IV, pp 1595–1610

Takashi N, Takeshi H, Masami Y, Masahiko B (2003) Preventing the escape of harmful elements using silica coating. J Japan Soc Eng Geol 43:390–395

Tao X, Wu P, Tang C, Liu H, Sun J (2012) Effect of acid mine drainage on a karst basin: a case study on the high-As coal mining area in Guizhou province, China. Environ Earth Sci 65:631–638

Tasse N, Germain D, Dufour C, Tremblay R (1997b) Hard-pan formation in the Canadian Malartic mine tailings: Implication for the reclamation of the abandoned impoundment. In: Proceeding of 4th international conference on acid rock drainage. Vancouver, BC, Canada, vol III, pp 1797–1812

Tasse N, Germain D, Dufour C, Tremblay R (1997a) Organic-waste cover over the eastern mine tailing: beyond the oxygen barrier. In: 4th conference on acid rock drainage, May 31–June 6, Voncouver, BC. Secretariat CANMET, Ottawa, Canada, vol 4, pp 1627–1642

Taylor JR, Guthrie B, Murphy NC, Waters J (2006) Alkalinity producing cover materials for providing sustained improvement in water quality from waste rock piles. In: Proceedings of the 7th international conference on acid rock drainage (ICARD), Mar 26–30, St. Louis, MO, American Society of Mining and Reclamation, Lexington, KY

Thompson JS, Jerkins RE (1999) Evaluation of the DuPont Passivation technology at the home stake, mine, lead, South Dakota (No. DC-JL-99-5). Jackson Laboratory, El Du Pont de Nemours

Timms GP, Bennett JW (2000) The effectiveness of covers at Rum Jungle after fifteen years. In: Proceedings 5th international conference on acid rock drainage. Society for mining, Metallurgy and Exploration, vol 2, pp 813–818

Vandiviere MM, Evangelou VP (1998) Comparative testing between conventional and microencapsulation approaches in controlling pyrite oxidation. J Geochem Explor 64:161–176

Vigneault B, Campbell PGC, Tessier A, Vitre RD (2001) Geochemical changes in sulfide mine tailings stored under a shallow water cover. Water Res 35:1066–1076

Waddell RK, Parizek RR, Buss DR (1980) The application of limestone and lime dust in the abatement of acidic drainage in centre county, Pennsylvania. Final report of research project 73-9. Commonwealth of Pennsylvania, Department of Transportation, Office of Research and Special Studies, 245p

Wang HL, Shang JQ, Kovac V, Ho KS (2006) Utilization of Atikokan coal fly ash in acid rock drainage from musselwhite mine tailings. Canadian Geotech J 43:229–243

Watzlaf GR, Erickson PM (1986) Topical amendments of coal refuse: effect on pore gas composition and water quality. In: Proceedings of national symposium on mining, hydrology, sedimentology, and reclamation, Lexington, KY, 8–11 Dec 1986, pp 225–261

Wu JT, Chiang YR, Huang WY, Jane WN (2006) Cytotoxic effects of free fatty acids on phytoplankton algae and cyanobacteria. Aquat Toxicol 80:338–345

Xenidis A, Mylona E, Paspaliaris I (2002) Potential use of lignite fly ash for the control of acid generation from sulfidic wastes. Waste Manag 22:631–641

Yanful EK (1993) Oxygen diffusion through soil cover on sulphidic mine tailings. J Geotech Eng 119:1207–1228

Yanful E, Orlandea M (2000) Controlling acid drainage in a pyritic mine waste rock. Part II: geochemistry of drainage. Water Air Soil Pollut 124:259–284

Yanful EK, Payant SC (1992) Evaluation of techniques for preventing acidic rock drainage. Milestone research report, MEND program, December

Yanful JM, Verma A (1999) Oxidation of flooded mine tailings due to resuspension. Can Geotech J 36:826–845

Yanful EK, Simms PH, Payant SC (1999) Soil cover for controlling acid generation in mine tailings: a laboratory evaluation of the physics and the geochemistry. Water Air Soil Pollut 114:347–375

Yanful EK, Orlandea MP, Eliasziw M (2000) Controlling acid drainage in a pyrite mine waste rock. Part I: statistical analysis of drainage data. Water Air Soil Pollut 122:369–388

Yanful EK, Oralandea MP (2000) Controlling acid driange in a pyritic mine waste rock. Part 2: geochemistry of drainage Water Air and Soil Pollution 124:259–284

Yeheyis MB, Shang JQ, Yanful EK (2009) Long-term evaluation of coal fly ash and mine tailing co-placement: a site specific study. J Environ Manag 91:237–244

Zaman KM, Chusuei C, Blue LY, Atwood DA (2007) Prevention of sulfide mineral leaching through covalent coating. Main Group Chem 6:167–184

Zhang YL, Evangelou VP (1996) Influence of iron oxide forming conditions on pyrite oxidation. Soil Sci 161:852–864

Zhang YL, Evangelou VP (1998) Formation of ferric hydroxide-silica coatings on pyrite and its oxidation behavior. Soil Sci 163:53–62

Zhang X, Borda MJ, Schoonen MAA, Strongin DR (2003a) Adsorption of phospholipids on pyrite and their effect on surface oxidation. Langmuir 19:8787–8792

Zhang X, Borda MJ, Schoonen MAA, Strongin DR (2003b) Pyrite oxidation inhibition by a cross-linked lipid coating. Geochem Transact 4:8–11

Zhang XV, Kendall TA, Hao J, Strongin DR, Schoonen MAA, Martin ST (2006) Physical structure of lipid layers of pyrite. Environ Sci Technol 40:1511–1515

Zhuping Z, Hecai N, Gerke HH, Huttl RF (1998) Pyrite oxidation related to pyrite mine site spoils and its controls: a review. Chin J Geochem 17:159–169

Urban vs. Rural Factors That Affect Adult Asthma

Yu Jie, Zaleha Md Isa, Xu Jie, Zhang Long Ju, and Noor Hassim Ismail

Contents

Y. Jie
Faculty of Medicine, Department of Community Health, National University of Malaysia,
Kuala Lumpur 56000, Malaysia

School of Public Health, Zunyi Medical College, Zunyi, Guizhou 563003,
People's Republic of China

Z.M. Isa • N.H. Ismail (✉)
Faculty of Medicine, Department of Community Health, National University of Malaysia,
Kuala Lumpur 56000, Malaysia
e-mail: hassim@ppukm.ukm.my

X. Jie
School of Public Health, Zunyi Medical College, Zunyi, Guizhou 563003,
People's Republic of China

Z.L. Ju
The First Department of Respiratory Diseases, First Affiliated Hospital of Zunyi Medical College,
Zunyi, Guizhou 563003, People's Republic of China

D.M. Whitacre (ed.), *Reviews of Environmental Contamination and Toxicology*
Volume 226, Reviews of Environmental Contamination and Toxicology 226,
DOI 10.1007/978-1-4614-6898-1_2, © Springer Science+Business Media New York 2013

1 Introduction

People in modern societies spend the vast majority of their time in indoor environments, including homes, workplaces, schools, and public spaces. Hence, indoor environmental quality has a significant impact on public health and well-being. Exposure or sensitization to indoor pollutants, including cigarette smoke (Hersoug et al. 2010), air pollution (Trupin et al. 2010), and allergens (Dottorini et al. 2007), is an important risk factor for asthma morbidity. Asthma, a common chronic respiratory disease, has been a growing international issue because its prevalence has been expanding in adults and children. The burden of this disease on governmental healthcare systems, patients and their families is increasing worldwide. It is estimated that there are approximately 300 million asthma patients worldwide and that 15 million disability-adjusted life years are lost annually by those afflicted with asthma (Fukutomi et al. 2010). Asthma usually arises from an interaction between host and environmental factors. A rapid increase in asthma in recent years cannot be ascribed to changes in genetic factors, but rather, to changes in environmental factors. In addition to increased indoor air contaminant exposures, several social factors that may contribute to developing asthma morbidity have been studied; among factors that have been given widespread attention are geographical variations, socioeconomic status (SES), and ethnicity.

It has been suggested that, compared to urban dwellers, people living in rural areas generally have better health, along with fewer disabilities and long-term limiting illnesses (Iversen et al. 2005). Moreover, rural residents smoke less, consume less alcohol, and are not as drug-dependent. Those residing in rural areas also have less psychiatric morbidity (Romans et al. 2011). Although they suffer poorer survival rates from lung or colorectal cancer, as compared to those living in urban areas, rural people encounter greater barriers to receiving healthcare services (Lu et al. 2010). Their access to care may be hindered by various factors, including inadequate insurance coverage, an undersupply of healthcare providers and facilities, lack of transportation and a proper healthcare policy where they reside. Coherent evidence shows that differences in the prevalence of asthma morbidity between urban and rural areas exists (Yemaneberhan et al. 1997; Ellison-Loschmann et al. 2004; Smith et al. 2009). Such difference in the prevalence of asthma morbidity may result from increasing urbanization, or from socioeconomic and cultural factors, as well as individual societal factors.

In Fig. 1, we present a diagram of our understanding of the human exposure pathways and potential factors that affect human asthma susceptibility between urban and rural areas. In this model, different environmental urban-based exposures (e.g., particulate or gaseous air pollutants from vehicular traffic, and industry), and similar ones in rural areas (e.g., indoor pollution from biomass fuel combustion, and keeping or herding animals) are known to potentially affect susceptible adult hosts. Such exposures may produce airway inflammation and obstruction. However, there

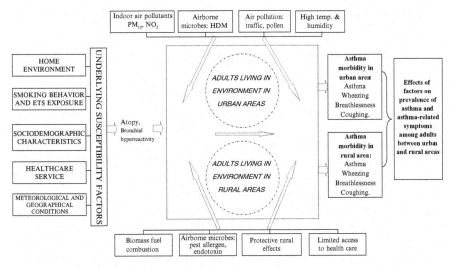

Fig. 1 A diagram of human exposure pathways and potential factors that affect adult asthma susceptibility between urban and rural areas (PM$_{10}$: particulate matter less than 10 microm in diameter; NO$_2$: nitrogen dioxide; ETS: environmental tobacco smoke; HDM: house dust mite; temp: temperature)

are disparities between urban and rural areas in the prevalence of asthma and asthma-related symptoms. Susceptibility factors are specific to urban and rural areas, and these differences potentially influence the incidence of asthma between the two areas. Factors that may be important in susceptibility to asthma include sociodemographic characteristics, type of home environment, availability of healthcare services, meteorological or geographical differences, and degree of urbanization. Furthermore, factors like atopy and specific genetic background of exposed persons influence host susceptibility to environmental stimuli.

Because asthma and asthma-related symptoms constitutes a growing health problem around the world (Yu et al. 2011a, b), we undertook to investigate the disparities that exist between those living in urban vs. rural areas, in regard to the prevalence of asthma and asthma symptoms that they incur. In particular, we evaluated the importance of geographic variations and their effects on asthma prevalence and morbidity among adults. In addition, we evaluated the possible causes of asthma morbidity between urban vs. rural residents.

This article extends our earlier work (Yu et al. 2011a, b), in which we tried to determine the factors that affect the prevalence of asthma, allergy, and respiratory symptoms among adults, and whether rural/urban living has an effect on asthma prevalence. Investigating the differences that environmental risk factors pose to adults on prevalence of asthma and asthma-related symptoms in urban and rural locations may provide clues to the mechanisms by which asthma and asthma morbidity occur, and may also uncover needs that are currently unmet.

2 Differences Between Urban and Rural Environments

2.1 Adult Asthma Prevalence and Morbidity in Rural vs. Urban Areas

The prevalence of asthma throughout China has markedly increased in recent years, having previously been uncommon. Different indoor environmental quality and housing, human characteristics between urban and rural areas are likely to cause important differences in exposure to indoor environmental risk factors. Geographical variation in the prevalence of asthma and asthma-related symptoms may also be closely related to discrepancies in sociodemographic characteristics and other risk factors. For example, in North America, wood and natural gas are sources of heating fuel that are used in rural areas, coal is less frequently used, and depends on what is most readily available in a particular geographical region.

Differences in adult asthma prevalence and morbidity between rural and urban communities have been investigated by many researchers (Ellison-Loschmann et al. 2004; Jackson et al. 2007; Smith et al. 2009; Gao et al. 2011; Musafiri et al. 2011; Frazier et al. 2012). Jackson et al. (2007) reported results of a 2000–2003 survey of lifetime asthma prevalence and trends in metro and nonmetro counties of the USA; they indicated that the prevalence of lifetime asthma diagnoses was 12% for metropolitan counties and 11% for nonmetropolitan counties. The incidence of lifetime asthma diagnoses trended upwards across the rural–urban spectrum between 2000 and 2003, and states with the highest 2003 prevalence and the greatest increase in prevalence among non-metropolitan residents were concentrated in the West Census region (e.g., Arizona and California). Asthma prevalence in non-metropolitan counties was highest for those aged 18–34 (15.9%), the unemployed (13.5%), American Indians (12.7%) and women (12.4%). The recommended approach for diagnosing and treating asthma, Jackson et al. (2007) concluded, may be more difficult to implement in rural counties. Smith et al. (2009) conducted one cross-sectional study in the USA and explored the risk factors associated with healthcare utilization among 3,013 Arizona Medicaid patients with asthma. Urban areas had higher rates of asthma-related hospital visits compared to rural counties, and adults had higher rates than adolescents, but there were no differences in asthma-costs between urban and rural areas. This suggested rural exacerbations may be more costly or severe. In Rwanda, Africa, Musafiri et al. (2011) investigated the prevalence of atopy and asthma in an urban and a rural area. The prevalence of asthma was higher in the urban than in the rural area. Risk factors for asthma included allergy (OR: 4.01; CI: 2.86–5.13), female gender (OR: 1.99; CI: 1.18–3.69) and living in a metropolis (OR: 3.62; CI: 2.12–5.78).

Gao et al. (2011) investigated the prevalence of asthma in China's Qinghai Province in 24,341 adults between 2006 and 2007. The prevalence of asthma in rural, urban, half-farming and half-herding areas, and in pastoral areas was 0.64% (65/10,119), 0.27% (37/13,933), 0.15% (2/1,310) and 0.04% (1/2,489) respectively; the highest incidence was in rural areas and the lowest was in the pastoral areas.

In New Zealand, Ellison-Loschmann et al. (2004) observed regional patterns of asthma hospitalizations in Maori vs. non-Maori areas. The rate of asthma hospitalization was higher for Maori than for non-Maori peoples in two different age-groups: 15–34 years RR = 1.31; and 35–74 years RR = 2.97. Moreover, the differences were higher in rural areas (RR: 1.34 and 3.13) than in urban areas (RR: 1.22 and 2.79) (Ellison-Loschmann et al. 2004).

Frazier et al. (2012), in another study, found that the prevalence of self-reported current asthma among adults is similar in metropolitan and nonmetropolitan counties in Montana of the USA, although other sociodemographic differences existed.

2.2 Combustion Products

2.2.1 Environmental Tobacco Smoke

Environmental tobacco smoke (ETS) contains and delivers over 4,000 compounds to those who are exposed to it, and among the components are carcinogenic agents such as benzo(a)pyrenes, polycyclic aromatic compounds, and tobacco-specific nitrosamines (Haverkos 2004). In 1993, the Environmental Protection Agency (EPA) classified ETS as a known human carcinogen and estimated that ETS exposure is responsible for approximately 3,000 lung cancer deaths each year among adult nonsmokers (USEPA 1993). Exposure to ETS has been associated with many detrimental health consequences, ranging from increased risk of asthma and asthma-related symptoms to lung cancer.

Over the past decades, positive associations between development of asthma and asthma attacks have been reported in children that have existing disease and exposure to ETS in the indoor environment. In contrast, data on the effects of ETS exposure on adults with asthma are limited. Hersoug et al. (2010) found that indoor exposure to ETS is associated with respiratory symptoms and diminished lung function in adults (Hersoug et al. 2010). Furthermore, in one recent 10-year cohort study, Polosa et al. (2011) revealed that cigarette smoking is an important predictor of asthma severity and poor asthma control (Polosa et al. 2011).

Exposures to ETS and Smoking Behavior in Urban vs. Rural Areas

The relationship between adverse health impacts and ETS may be influenced by a disparity of geographic location (Ho et al. 2010). Collected evidence suggests that those residing in rural areas may be disproportionately affected by the health burden of tobacco use. For example, although China has made progress towards achieving a smoke-free environment, there remains a high degree of exposure to secondhand smoke. Xiao and colleagues (2010) have recently studied Chinese nonsmokers aged 15 years and older ($n = 13,354$) to determine the extent of their secondhand smoke exposure. Results showed that 72.4% (556 million) were exposed to secondhand smoke, with 52.5% (292 million) exposed to secondhand smoke daily. The prevalence of secondhand smoke (SHS) exposure was 70.5% for urban populations, and 74.2% for rural populations, respectively (Xiao et al. 2010). In addition, in another behavioral risk factor survey in the USA, Vander Weg et al. (2011) suggested that adults residing in rural areas were significantly more likely to smoke cigarettes, 22.2% vs. 17.3% (suburban) and 18.1% (urban); the exposure disparity resulted from both their own tobacco use and exposure to others' cigarette smoke (Vander Weg et al. 2011).

Differences in Prevalence of Asthma and Asthma Morbidity Among Adult Smokers in Urban vs. Rural Areas

There is conflicting evidence about the risk of asthma and asthma-related symptoms in adults living in rural areas, compared to those living in urban areas (Woods et al. 2000; Zhang et al. 2002; Ghosh et al. 2008, 2009; Lâm et al. 2011). Two cross-sectional studies were performed that indicated the prevalence of asthma morbidity was higher among rural smokers (Woods et al. 2000; Ghosh et al. 2009), and another two suggested the opposite (Ghosh et al. 2008; Lâm et al. 2011). A survey of 11,223 Canadian adults found that rural/urban locations modified the relationship of asthma prevalence and smoking, but only among women. Female smokers and ex-smokers in rural areas were 1.4 times (95% CI: rural smokers = 1.02–1.94, and rural ex-smokers = 1.02–2.02) more likely to be diagnosed with asthma, in contrast to non-smoking urban women. The reason that rural locations influence effects may result from females having more sensitive airways, may exacerbate respiratory symptoms from smoking and may have higher exposure to fumes and dusts in rural locations (Ghosh et al. 2009). A cross-sectional study of randomly selected young adult inhabitants (20–44 years old) in Australia produced similar results in south-eastern metropolitan Melbourne ($n=4,455$) and rural south-western New South Wales ($n=4,521$). Respondents from the Riverina in New South Wales reported a significantly higher prevalence of nocturnal dyspnea, chronic bronchitis, and asthma attacks in the previous 12 months, or ever having had asthma and doctor-diagnosed asthma, compared to those from Melbourne ($p<0.05$). Riverina respondents reported a higher prevalence of smoking ($p<0.05$) and smoked more cigarettes on average ($p<0.001$) than Melbourne respondents (Woods et al. 2000).

By contrast, a cross-sectional study using the Canadian National Population Health data, collected from 1994 to 2000, showed that the prevalence of asthma among smokers and nonsmokers were more prominent in urban than in rural residents. Higher stress levels and lack of open spaces, compared with their rural counterparts, may be possible reasons for the higher asthma prevalence among smokers living in urban areas. Environmental factors and exposure to secondhand smoke may be possible reasons for the higher asthma prevalence among nonsmokers in urban areas (Ghosh et al. 2008). In Vietnam, 3,008 subjects living in an inner city area of Hanoi and 4,000 in a rural area of Bavi, in northern Vietnam, were studied. Results show that the prevalence of asthma in adults may have increased in both urban and rural Vietnam. Nearly One half (49.7%) of the men living in the inner city area of Hanoi were smokers, as were a majority (67.8%) living in a rural area of

Bavi; by contrast, only a few percent of women in both areas were smokers. The prevalence of ever having had asthma in Hanoi was 5.6% (Bavi 3.9%; $p=0.003$), with no major gender difference (Lâm et al. 2011).

The results of one study (Walraven et al. 2001) were inconsistent with others, because the study failed to link tobacco smoking to the onset of asthma in adults. These authors reported findings of a large study performed among 2,166 participants in the urban population vs. 3,223 participants in the rural population in Gambia. In the urban population, 4.1% reported asthma-related symptoms, 3.6% reported doctor-diagnosed asthma, whereas in the rural population, these figures were 3.3% and 0.7%, respectively. Both male and female participants who lived in rural locations (42%) smoked more than those who lived in urban areas (34%). Notwithstanding that, the risk of asthma was not elevated in active smokers compared with never-smokers.

2.2.2 Coal and Biomass Fuels

Almost half the world's population, and up to 90% of rural households in developing countries, still use solid fuels such as coal, firewood, wood chips, crop residue, and dung cakes for their domestic energy needs (Bruce et al. 2000). Indoor combustion of coal or biomass fuels produces both gases and particulate matter that can affect the development and exacerbation of asthma. The best understood of these substances are particulate matter, carbon monoxide, sulfur oxides, nitrous oxides, volatile organic compounds (e.g., formaldehyde), and polycyclic organic matter such as benzo[a]pyrene.

Particulate Matter

It has been documented in epidemiologic studies that there is a relationship between air pollution from indoor combustion sources and asthma and asthma-related symptoms (Qian et al. 2007; Mishra 2003). Although such collective evidence supports the role of particulate matter (PM) exposure in the development of asthma, and as a potential risk factor of asthma exacerbations, the mechanisms by which cooking fuel smoke (mainly particulate matter) influence asthma are not well understood. It has been postulated that particulate matter may induce pulmonary health effects through oxidative stress and pro-inflammatory effects, which have been documented to occur in both in vivo and in vitro studies (Lei et al. 2004; Riva et al. 2011).

Several studies have shown that pollutants from coal and biomass fuel combustion can be present in environment of rural and urban areas (Mestl et al. 2007a; Jiang and Bell 2008; Fullerton et al. 2009; Colbeck et al. 2010a, b). In a recent study from Pakistan, Colbeck et al. (2010a) found high indoor levels of PM during cooking, with concentrations in the range of 4,000–8,555 $\mu g/m^3$. The average indoor–outdoor ratios for PM_{10} (1.74), $PM_{2.5}$ (2.49), and PM_1 (3.01) in the living rooms of a rural home vs. an urban one corresponded to the ratios of PM_{10} (1.71), $PM_{2.5}$ (2.88), and PM_1 (3.47), respectively (Colbeck et al. 2010a).

Approximately, 3.1 million annual deaths in rural China are attributable to indoor air pollution (IAP) (Mestl et al. 2007b). Mestl et al. (2007a, b) reported the largest exposure burden to be in counties that rely heavily on biomass for energy, such as China does. The average exposure is estimated at 340 ± 55 $\mu g/m^3$ in southern Chinese cities, and 440 ± 40 $\mu g/m^3$ in northern ones, whereas the average rural population exposure was 750 ± 100 $\mu g/m^3$ and 680 ± 65 $\mu g/m^3$ in the south and north, respectively (Mestl et al. 2007a). Furthermore, in another study from China, in which the PM_{10} levels from biomass-burning in rural vs. non-biomass-burning urban households were compared, the monitoring results showed rural kitchen PM_{10} levels (202.1 ± 293.6 $\mu g/m^3$) to be three times higher than those in urban kitchens (67.00 ± 32.58 $\mu g/m^3$) during cooking. The highest $PM_{2.5}$ exposures occurred during cooking periods for urban and rural cooks. However, rural cooks (487.9 ± 874.9 $\mu g/m^3$) had 5.4 times higher $PM_{2.5}$ levels while cooking than did urban cooks (90.1 ± 120.9 $\mu g/m^3$) (Jiang and Bell 2008).

Few authors, to date, have evaluated the relationship between adult asthma and asthma-related symptoms and domestic fuel combustion exposures in houses between urban and rural areas. A survey of 3,709 Chinese adults in Beijing, Anqing City, and rural communities in Anqing Prefecture disclosed significant differences between study areas in the prevalence of chronic cough, chronic phlegm, wheezing and dyspnea ($p < 0.05$). The lowest prevalence of respiratory symptoms was observed in Anqing City, a higher prevalence occurred in rural Anqing, and the highest prevalence was observed in Beijing. In this study, median indoor concentrations of PM_{10} were similar in Anqing City (239 $\mu g/m^3$) and rural Anqing (248 $\mu g/m^3$), but were much higher in Beijing (557 $\mu g/m^3$), while median indoor concentrations of SO_2 were similar in all three areas (Venners et al. 2001). The measured indoor PM_{10} concentrations in this study were far higher than the American Society of Heating, Refrigerating & Air-Conditioning Engineer (ASHRAE)'s standards for indoor 24-h average PM_{10} of 150 $\mu g/m^3$ (ASHRAE 1989).

Although asthma is less common among rural populations where biomass fuels are in common use, compared with their urban counterparts, it should not be assumed that smoke exposure is not deleterious in these areas. The relationship between residential exposure to solid biomass fuels (animal dung, crop residue, and wood) and asthma morbidity was also examined in Nepal, in which a cross-sectional study ($n = 168$) of a representative sample of housewives was carried out (Shrestha and Shrestha 2005). The prevalence of asthma morbidity, including cough, phlegm, breathlessness, wheezing and asthma were higher among those living in mud and brick houses, compared to concrete houses, and the prevalence was also higher among residents living on hills and in rural areas, in contrast to those living on flatland and in urban areas.

Nitrogen Dioxide

Nitrogen dioxide (NO_2), a combustion by-product, is predominantly an indoor pollutant, and the major source of exposure to it is from household appliances fuelled by gas, particularly in households without flues for such gas appliances (Spengler et al. 1979). Inhaling high concentrations (>5 ppm) of NO_2 is associated with acute epithelial damage, whereas exposure to lower concentrations that exist in urban areas or in an indoor environment can aggravate preexisting lung disease, including asthma (Persinger et al. 2001). To date, the relationships between asthma and asthma-related outcomes and indoor nitrogen dioxide exposures in urban vs. rural areas have rarely been investigated. The results of a recent study, in which the levels of NO_2 were assessed at a Pakistani urban and two rural sites were interesting. In winter, NO_2 concentrations at all three sites were higher in kitchens than living rooms or outdoors. During the summer, NO_2 levels fell sharply at both rural sites (respectively from 256 $\mu g/m^3$ and 242 $\mu g/m^3$ to 51 $\mu g/m^3$ and 81 $\mu g/m^3$). However, at the urban site, the mean levels were slightly higher in summer (234 $\mu g/m^3$) than in winter (218 $\mu g/m^3$). Elevated NO_2 concentrations may pose a significant threat to health and especially to the vulnerable population (e.g., women and children) (Colbeck et al. 2010b).

The mechanisms by which inhaled NO_2 adversely affect human health are difficult to delineate. Sandstrom et al. (1992) reported that repeated exposure to 4 ppm of NO_2 may have adverse implication on local bronchial immunity by reducing total macrophages, B cells, NK lymphocytes, peripheral blood lymphocytes and by reducing the T-helper-inducer/T-cytotoxic-suppressor ratio in alveolar lavage (Sandstrom et al. 1992). In another study, it was demonstrated that NO_2 induced bronchial inflammation and a minimal change in bronchoalveolar lavage T-helper cells (Solomon et al. 2000). Prior studies suggest nitrous acid (HONO), a molecule that can be formed as a primary product of gas combustion or by the reaction of NO_2 with surface water, may contribute to adverse health outcomes previously attributed to NO_2 (Jarvis et al. 2005). The theoretical health risks of HONO include damage to the mucous membranes and lungs by direct contact with the acid, creation of carcinogenic nitrosamines secondary to the HONO combination with amines, and oxygen-free radical production through HONO photolysis in air (Peterson. 2012).

2.3 Biological Contaminants

Biological contaminants include a wide variety of biological agents commonly found in indoor environments. These contaminants include the following: (1) viruses; (2) bacteria (and any related endotoxins); (3) allergens, including house dust mite allergens [*Dermatophagoides farinae* 1 (Der f 1), *Dermatophagoides pteronyssinus* (Der p 1)], and allergens from animal dander [cat (*Felis domesticus* 1 (Fel d 1)), dog (*Canis familiaris* 1(Can f 1)), cockroach (*Blattella germanica* 1 (Bla g 1 & 2)), and mouse (*Mus musculus* 1 (Mus m 1)), mouse urinary protein (MUP) allergens], and fungi (including associated allergens, toxins, and irritants) (Yu et al. 2011a, b). Exposure to indoor allergens and molds, together with building dampness, is an important risk factor for the asthma morbidity of the occupants. However, the role of indoor allergen sensitization in contributing to asthma and asthma-related symptoms, among adults, remains a subject of controversy.

Geographic locations may have a predominant impact on the course of spore occurrence, biological contaminant levels, as well as on Seasonal Fungal Index (SFI) values. Biological contaminant levels and exposure characteristics in urban and rural areas have been compared in various studies (Taksey and Craig 2001; Armentia et al. 2002; Loureiro et al. 2005; El-Shazly et al. 2006; Kasprzyk and Worek 2006; Guinea et al. 2006; Oliveira et al. 2009).

2.3.1 Differences in Exposure to Fungal Allergens in Urban vs. Rural Areas

Several studies have suggested that exposure to fungal allergens could be different between urban and rural settings. The authors of one study in Poland's urban/rural areas found several spore types, including Alternaria, Botrytis, Cladosporium, Epiccocum, Ganoderma, Pithomyces, Polythrincium, Stemphylium, Torula, and Drechslera, that were significantly higher in rural than that in urban environments. Different habitat characteristics such as urbanization level, vegetation, and micro-climate seem to play an important role in determining the composition and concentration of the fungal airborne spore population in urban and rural areas (Kasprzyk and Worek 2006). In Spain, Guinea et al. (2006) observed that, despite the different sampling techniques employed, Aspergillus/Penicillium spore concentrations were more frequent in urban than in rural environments in Madrid.

Oliveira et al. (2009) investigated allergenic airborne fungal spores in urban and rural areas of Portugal. The mean daily spore concentrations in the rural and urban areas were 934 and 531 spores/m^3, respectively. The most abundant fungal spore types in rural and urban locations were Cladosporium (59.5% vs. 62.7%), Agaricus (5.6% vs. 4.3%), Agrocybe (1.2% vs. 1.4%), Alternaria (1.1% vs. 1.0%), and Aspergillus/Penicillium (0.9% vs. 1.1%) (Oliveira et al. 2009). In the USA, Taksey and Craig (2001) found that different allergen levels do exist in different regions of the country. Allergies to dust mites, dogs, timothy grass, and ragweed are more often reported by residents in rural than in large metropolitan centers. In contrast, cockroach and Alternaria hypersensitivity were more common in large cities (Taksey and Craig 2001).

2.3.2 Differences in Exposure to Endotoxins in Urban vs. Rural Areas

Asthma and asthma-related symptoms are highly prevalent in both urban and rural areas, and the prevalence of such symptoms has been closely related to bacterial endotoxins. Endotoxin levels have been studied in residential environments in urban and rural communities. In the USA, Roy et al. (2003) quantitified the bacterial DNA and endotoxin content in dust in urban vs. rural areas of Mississippi. Dust endotoxin levels in farm homes (mean, 22.1 mg/g dust), rural homes (6.3 mg/g), and farm barns (2.2 mg/g) are much higher than those in urban and nonfarm homes (0.6 mg/g) respectively (Roy et al. 2003). In addition, asthma hospitalization rates were significantly higher among all demographic groups in the rural (Mississippi) Delta region compared with the urban Jackson Metropolitan Statistical Area ($p < 0.001$) (Roy et al. 2010).

2.3.3 Difference in the Prevalence of Allergic Diseases After Exposure to Biological Contaminants

Variability in urban–rural prevalence of asthma and asthma-related symptoms has been observed in many parts of the world (Nguyen et al. 2010). Several authors of international studies have reported that exposure of sensitized asthmatics to allergens affected asthma prevalence and morbidity in urban vs. rural areas.

Residents living in urban areas are more likely to experience asthma respiratory symptoms than those living in rural areas (Eduard et al. 2004; Douwes et al. 2007). The relationship between geographic locations and prevalence of allergic diseases was examined by monitoring trends and determinants of cardiovascular disease in a European project. This project was called Multinational Monitoring of Trends and Determinants in Cardiovascular Disease (MONICA), and was a cross-sectional study ($n = 664$) of a representative sample of adult aged 25–75 years in Germany (Filipiak et al. 2001). Results were that, apart from asthma and sensitization against house dust mite, significant differences in risk of allergic rhinitis (OR. 1.5; 95% CI: 1.2–1.9), atopic sensitization (OR. 1.2; 95% CI: 1.0–1.4) and sensitization against pollen (OR. 1.5; 95% CI: 1.2–1.9) were found in urban vs. rural residents. Given that protective rural effects exist in allergic rhinitis, farmers may experience lower risks in allergic rhinitis, atopic sensitization, and sensitization against pollen and mites, in contrast to rural non-farming residents.

In a cross-sectional study performed with 55 patients in Poland, the differences in clinical test results were examined for patients suffering bronchial asthma. Some patients lived in rural, and some in urban areas. In this study, significant ($p < 0.05$) disturbances in respiratory parameters were found in 23% of patients living in the city, and in 50% of subjects from the rural area. The level of IgE against grass pollens was significantly ($p < 0.01$) higher in urban residents. The results showed important differences between clinical manifestation of asthma in rural and urban patients, and indicated that a difference in etiology may play an important role in asthma development (Krawczyk et al. 2003).

In contrast to other studies, one multicentre study from Turkey found a direct association between allergic diseases and visible mold at home. Living in a house with visible mold increased the risk of respiratory illnesses such as asthma and wheezing, especially in rural areas (Kurt et al. 2009).

Migration from rural to urban areas also may increase the risk of allergy. In Denmark, a prospective population-based study performed over an 8-year observation period showed that adults who had migrated from rural to an urban area (Copenhagen) were at increased risk of developing an allergy (OR: 3.4, 95% CI, 1.6–7.2). This finding showed that persons migrating from rural to urban areas may acquire some new risk factors in the urban environment, or may lose some protective factors resulting from life in the rural environment, thus decreasing immune tolerance and increasing allergy risk (Linneberg 2005).

2.3.4 Protective Effects of Rural Living on Development of Asthma and Asthma Morbidity

Previous studies have disclosed that people living in urban- or Western-lifestyle countries appear to be subject to higher asthma mortality. Exposure to livestock was increasingly associated with a decreased burden of asthma-related diseases (Ghosh et al. 2008). Protective factors were associated with farm milk operations that started very early in life (von Mutius and Vercelli 2010), rural ways of living (Viinanen et al. 2007; Anastassakis et al. 2010), and having had early childhood infections (e.g., parasitic infections) (Yemaneberhan et al. 1997; Nyan et al. 2001), as well as bacterial exposure from animals in rural settings (Senthilselvan et al. 2003). The mechanism for this is unclear. It was pointed out in early studies that the immune system, when exposed to high levels of allergens, or children when exposed early to infections, may induce forms of immune tolerance or immunological response; it is such events that reduce the risk of developing asthma (von Mutius et al. 2000; Busse and Lemanske 2001).

Numerous studies have shown that rural living is protective against asthma and asthma morbidity. To assess whether children who grow up in a farming environment have been protected against a general increase in atopic disorders in Sweden, Bråbäck et al. (2004) used the Swedish Military Service Conscription Register, the Register of the Total Population and the Population and Housing Censuses data to conclude and report the following: the adjusted risk ratios for asthma in conscripts from farming vs. non-farming families were 1.00 (95% CI, 0.93–1.07), 0.94 (95% CI, 0.88–1.01), and 0.85 (95% CI, 0.79–0.91) in conscripts born in 1952–1961, 1962–1971, and 1972–1981, respectively. An inverse association was observed between farm living and asthma. The protective effect of growing up on a farm on the risk of asthma appears to be a fairly recent phenomenon (Bråbäck et al. 2004).

It is possible that the farming environment and rural lifestyle may be associated with an unknown "protective farming" effect, in terms of the disparity of prevalence of asthma morbidity in urban vs. rural communities. In another cross-sectional questionnaire survey of 9,453 Mongolian residents (Viinanen et al. 2005), a subgroup of 869 underwent skin prick tests, spirometry and bronchodilation, or methacholine-challenge testing. In this subgroup, the prevalence of asthma, allergic rhinoconjunctivitis, and allergic sensitization (with 95% CI) were 1.1% (0.3–2.0%), 9.3% (4.0–14.6%), and 13.6% (7.4–19.9%) in Mongolian villages, 2.4% (1.4–3.5%), 12.9% (8.2–17.7%), and 25.3% (17.1–33.6%) in rural towns and 2.1% (1.3–3.0%), 18.4% (13.3–23.4%), and 31.0% (24.5–37.5%) in Ulaanbaatar city. The authors implied that the prevalence of atopic diseases was lower in rural than in suburban or urban areas. Moreover, rural living environments (e.g., presence of herd animals, fermented milk products exposure) protected against atopy (Viinanen et al. 2005).

Furthermore, in a large study in Scotland, Iversen et al. (2005) investigated rural/urban differences in the prevalence of self-reported asthma and asthma-related symptoms. In this study, a significantly lower prevalence of asthma (adjusted OR, 0.59; 95% CI, 0.46–0.76) and eczema (adjusted OR, 0.67; 95% CI, 0.52–0.87) were reported in rural than urban areas, as were the prevalence of persistent cough, phlegm, breathlessness, and wheezing, although no cause and effect relationship was found. The investigators inferred that rural residency may be associated with better health status because of the protective role afforded by the presence of livestock against allergy among subjects (Iversen et al. 2005).

Another cross-sectional study was performed on university students from Finland who spent their childhoods either on a farm or in a nonfarm environment (Kilpeläinen et al. 2001). In this study, a total of 10,667 subjects aged 18–25 years were recruited. Current asthma was found to exist in 3.1% of subjects having lived childhoods in a farm environment; the corresponding number for those living in a nonfarm environment was 12.4% (OR: 0.22; 95% CI, 0.07–0.70). A more moderate or severe bronchial hyperreactivity (methacholine $PD_{20} FEV_1 \leq 600$ μg) was found among subjects who resided as children in a nonfarm environment. Cat-specific IgE was significantly negatively associated with a childhood in a farm environment (1.5% vs. 13.1%; OR: 0.10, 95% CI, 0.02–0.47) (Kilpeläinen et al. 2001).

In addition, Raukas-Kivioja et al. (2007) reported a study of 1,346 Estonia adults aged 17–69 years, who lived in urban or suburban areas before the age of 5, and had a significantly increased risk for allergic sensitization compared to those who were rural residents. In this study, the most common sensitizer in urban areas was cockroach allergens; whereas, in rural areas, the primary sensitizer was storage mites. Positive skin prick test (SPT) results were found in 27% vs. 40% of the study subjects who had lived in a rural vs. urban environment during early childhood; the highest prevalence was found in the most heavily polluted urban areas (Raukas-Kivioja et al. 2007).

In previous studies that compared asthma prevalence between urban and rural settings, all findings indicated decreased prevalence of asthma in rural areas. These results may result from the higher beneficial exposure of rural populations to farming in general and to endotoxins (Valet et al. 2009). The role of endotoxins in asthma is somewhat paradoxic for how they affect children and adults in rural environments. Liu (2004) presumed that asthmatics are particularly sensitive to inhaled endotoxin, and inhalation induces both immediate and sustained airflow obstruction. The paradox of endotoxin exposure is that higher levels of exposure early in life may mitigate the later development of allergy and persistent asthma. With endotoxin exposure being significantly higher in homes with animals and in farming households, where allergy and asthma are less likely to develop, endotoxin and other microbial exposures in early life may keep allergen sensitization and asthma from developing by promoting Th1-type immune development (Liu 2004).

2.3.5 Atopy Associated with Asthma Among Adults Living in Rural and Urban Areas

The underlying factors that affect the course of asthma are not well understood in either rural or urban settings. Additionally, the interactions between atopy and environmental exposure in modulating asthma have not been investigated intensively. Three scientific articles were identified that associated atopy and asthma morbidity in urban vs. rural areas. Ige et al. (2011) recently reported that atopy is related to adult asthma in rural and urban communities in southwest Nigeria. Asthmatics in both urban and rural settings had significantly higher positive skin reactions to house dust mite, cockroach, mold, and mouse epithelium than did non-asthmatic controls ($p < 0.05$). Mean total serum IgE was also significantly higher in asthmatics than in nonasthmatic controls (Ige et al. 2011). Another study performed by Lourenço et al. (2009) in Portugal demonstrated that asthma was more frequently associated with nonallergic rhinitis (NAR) than with allergic rhinitis (AR) among adults. Although sensitization profiles were not different between the urban and rural patients, AR prevalence was higher in urban asthmatic patients than in rural asthmatics (77.3 vs. 68.2; $p = 0.009$). Grass pollen and mites were the major sensitizers for allergic rhinitis patients (Lourenço et al. 2009). Aggarwal et al. (2006) studied 73,605 respondents in urban and rural areas in Chandigarh, Delhi, Kanpur, and Bangalore, India, and found that the overall prevalence of asthma was 2.38%. Residing in an urban area and having a history suggestive of atopy combined with other factors (e.g., the female gender, advancing age, and lower SES) were associated with significantly higher odds of having asthma (Aggarwal et al. 2006).

2.4 Air Pollution and Geographical and Meteorological Conditions

2.4.1 Air Pollution

Outdoor air quality has a lesser or greater degree of influence on indoor air quality, depending on air exchange rates and the pollutant type involved. The major sources of air pollutants in the air of urban areas result from coal combustion, urban traffic emissions, suspended building construction dusts and the chemical industry. Moreover, there has recently been a rapid increase in the number of motor vehicles in some communities, especially in developing countries (e.g., the number of motor vehicles in Guangzhou increased 3.71-fold from 97,200 in 2002 to 555,200 in 2006; Guangzhou Statistics Yearbook 2007). This explosion of car ownership has produced traffic gridlock along with a proliferation of auto fumes, as the city's reliance on cars and trucks leaves its air with few reprieves.

People who live in urban areas tend to be more affected by asthma, allergy, and respiratory diseases than those who live in rural areas. Individuals exposed to higher levels of vehicular traffic have a higher frequency of asthma morbidity than those less exposed to traffic, which may explain some differences for those living in these two areas. Numerous authors have reported significant health impacts associated with exposure to air pollutants (Mugusi et al. 2004; Smith et al. 2009). Although the mechanism by which air pollution affects the development of asthma and asthma-related symptoms is unclear, increasing evidence has shown that the elevated indoor/outdoor levels of aeropollutants (e.g., particles from vehicular traffic) may enhance the severity of asthma morbidity in urban and rural areas.

Few measurements of aeropollutants in urban and rural areas have been reported in the scientific literature. Matson (2005) reported that the concentration of ultrafine particles from air pollution was higher in the metropolitan area of Copenhagen than in the medium-sized city of Gothenburg, and was lowest in more rural sites (Matson 2005).

The relationship between asthma prevalence and morbidity and ambient air pollution was studied in several previously published studies. In Australia, Woods et al. (2000) reported that among residents who had asthmatic attacks in the previous 12 those in Melbourne had a higher frequency of attacks than did residents of Riverina (a rural area). The impact from air pollution on the local population was greater in Melbourne than in Riverina (Woods et al. 2000). By contrast, one case–control study from Sweden explored 203 adult human cases, and 203 controls (aged 20–60 years), for differences in risk of asthma from traffic-related air pollution exposure at home. Differences were then compared between those living in urban and rural areas. Living in a rural area tended to increase the risk of developing asthma, when traffic pollution was adjusted for. Rural living may, therefore, be seen as an indicator of a certain lifestyle and of different exposure patterns (Modig et al. 2006).

Michnar et al. (2003) studied the incidence of allergic respiratory diseases between urban and rural areas against the environmental pollution level in Poland. They distributed a standard questionnaire to 1,223 adults living in randomly selected districts (Nałeczów, Puławy, Motycz-Lublin) of that country, wherein various levels of industrial pollution existed. No difference in bronchial asthma incidence was found in any of the districts, but the diagnosis rate of allergic rhinitis was higher in rural than in urban areas (Michnar et al. 2003).

To evaluate the extent to which climate and outdoor NO_2 pollution can explain the geographical variation in the prevalence of asthma and allergic rhinitis, a questionnaire-based survey of 18,873 adults (aged 20–44 years) was performed in Italy (de Marco et al. 2002). Those surveyed lived in 13 areas of northern Italy that enjoyed sub-continental and Mediterranean climates. Those living in the Mediterranean climate areas, characterized by higher annual mean temperature (16.2 °C vs. 12.9 °C) and lower NO_2 levels (31.46 µg/m^3 vs. 57.99 µg/m^3), had a significantly higher prevalence of asthma morbidity ($p < 0.001$) than subcontinental ones did. The investigators presumed that air pollution from traffic emissions may pose long-term adverse effects on adult pulmonary health such as asthma morbidity.

The World Health Organization cautiously concluded that traffic-related air pollution may enhance asthma development and exacerbate asthma-related symptoms, particularly among susceptible subgroups (Heinrich and Wichmann 2004). Nevertheless, air pollution may not be a major risk factor for developing asthma; rather, it is merely a minor trigger in some individuals. For example, in some regions (e.g., China and Eastern Europe) that have a high concentration of some air pollutants (e.g., PM_{10} and SO_2), there is generally a low rate of asthma prevalence. In contrast, some regions that have low levels of air pollution, such as some parts of New Zealand, still experienced a high prevalence of asthma. This suggests that even though it is plausible that outdoor air pollution plays a role in the increasing prevalence of asthma morbidity, other causes exist.

Study results from North America (Arif et al. 2004), Central America (Cooper et al. 2009), Europe (Filipiak et al. 2001), Asia (Viinanen et al. 2005), and Africa (Musafiri et al. 2011) have shown that the prevalence of asthma, and the morbidity associated therewith, is higher in urban than in rural environments. This is particularly true for pollinosis, whereas pollen counts are usually higher in urban than in rural areas (Armentia et al. 2002; Bousquet et al. 2008).

It has been demonstrated in several studies that plant pollen is present in both urban and rural environments. Armentia et al. (2002) reported that the protein content and allergenicity of *Lolium perenne* pollen was higher in urban than in rural areas. Such differences may explain why allergies to grass pollen more frequently occur in urban areas (Armentia et al. 2002). In comparing the distribution of pollen in urban and rural areas across an urbanization gradient, Bosch-Cano et al. (2011) found that the pollen burden from grass, ash, birch, alder, hornbeam, hazel and plantain exceeded the allergy threshold more often in rural than in urban settings, whereas, in urban areas, only plant-pollen quantities exceeded the allergy threshold more often than in rural areas (Bosch-Cano et al. 2011).

Asthma and asthma-related symptoms are common problems in both children and adults, but the causative pollen allergens vary by geographical area. Few academic studies have enumerated the correlation of asthma and/or allergen sensitivity with geography among adults. Loureiro et al. (2005) revealed, in a large study of 1,096 patients from Cova da Beira, Portugal, the frequency of aeroallergens sensitization. They compared the sensitization in urban vs. rural environments. Respectively, results of urban–rural sensitivity to various allergenic sources were as follows: *D. pteronyssinus* (32% vs. 34.7%), mold mixtures (15.3% vs.12%), cat dander (17.1% vs. 15.2%), grass mixtures (51.3% vs. 36.4%), and *Parietaria judaica* (29.4% vs. 14%). The sensitization to indoor aeroallergens was lower than that to pollens. The authors assumed that pollution-enhanced sensitization to pollens occurred in the urban environment (Loureiro et al. 2005). Moreover, in Germany, Filipiak et al. (2001) reported that allergic rhinitis from sensitization to pollen was 2.5 times (95% CI: 1.8–3.6) more prevalent in urban than in rural residents (Filipiak et al. 2001).

2.4.2 Climatic and Geographic Factors

A potential role of various seasonal and meteorological factors in the etiology of asthma and asthma-related symptoms has long been suspected. There is evidence that climate may affect asthma symptom prevalence and frequency, either directly (e.g., via an effect of air temperature on airway responsiveness) or indirectly (e.g., via altered exposure to infections, aeroallergens, or air pollutants) (Metintas and Kurt 2010; Franco et al. 2009; Hales et al. 1998; Verlato et al. 2002; Chen et al. 2006).

Different geographical and climatic properties (e.g., elevation, temperature, humidity) have a close relationship with asthma morbidity prevalence, such as allergic rhinitis among adults in urban and rural areas. Many geographic locations have climates that are favorable to indoor mold growth, for example, in tropical areas that routinely have high humidity.

In a national survey conducted in five geographic regions in Thailand, Dejsomritrutai et al. (2006) found the prevalence of current diagnosed asthma in northeastern rural regions (2.05%) to be more prevalent than in central urban regions (1.66%). The higher temperatures and rainfall, together with higher incomes of residents of central regions vs. those living in northeast regions, were thought to account for differences in the prevalence of asthma in Thailand (Dejsomritrutai et al. 2006). Moreover, Lewis et al. (1997) studied 31,470 adults in New Zealand and found that a high frequency of asthma symptoms were experienced by New Zealand adults (15.2%); Maoris (22.1%) and women (17.0%) experienced a higher rate of such symptoms. This study also shows significant urban–rural differences, as well as marked differences in prevalence among various rural areas (Lewis et al. 1997).

2.5   Socioeconomic Characteristics and Healthcare Services

Recent evidence suggests that prevalence of asthma and rates of exacerbation (e.g., asthma attacks and related emergency room visits and hospitalizations) may, in fact, be higher than previously thought (Yu et al. 2011a, b; Comhair et al. 2011; Nguyen et al. 2011). In addition, such outcomes, when they occur, are associated with geographic and socioeconomic circumstances. Identifying socioeconomic and regional factors that contribute to prevalence of asthma, allergy, and respiratory symptoms among adults is complicated. Such complications result from the influence of multiple demographic indicators, particularly race and income, and of system-level considerations such as hospital admission and prescription policies that may vary at the local level. In addition, poor SES may confound the relationship between indoor climate and respiratory disease, because SES is associated with both the exposure and the outcome variable (Skorge et al. 2005).

Many studies have shown that asthma continues to disproportionately affect minority and low-income groups. SES factors, such as cultural attitudes toward health (Ellison-Loschmann et al. 2004), low levels of education and income (Joshi et al. 2003), lack of healthcare coverage (Ansari et al. 2003), and poorly maintained housing, as well as poor hygiene conditions (e.g., cockroach, rodent, etc.) (von Mutius and Vercelli 2010), which are the typical characteristics of rural areas, may contribute to the differences in the prevalence of asthma morbidity between urban and rural areas. Using the 2005 Behavioral Risk Factor Surveillance System (BRFSS) data from the USA, Morrison and coworkers (2009) investigated adult sociodemographic and health behavior contributions to asthma prevalence. After adjusting for sociodemographic and behavior characteristics, results were that no significant difference existed for residents of urban vs. rural areas. However, asthma prevalence in metropolitan regions was less pronounced than those in remote areas (OR: 0.96; 95% CI, 0.90–1.02) (Morrison et al. 2009). In contrast, in a national study in Australia, Cunningham (2010) reported asthma prevalence was higher for indigenous Australians than for nonindigenous Australians. The age and sex-adjusted relative odds of contacting asthma was 1.2 (95% CI, 1.0–1.5) in the nonindigenous population vs. 1.0 (95% CI, 0.8–1.3) in the indigenous population.

The socioeconomic pattern of those contacting asthma among indigenous Australians is much less clear. This may be due, in part, to the episodic nature of asthma, and to the limited health literacy and/or limited access to health care of this group. It may also reflect the importance of exposures to socioeconomic effects among the indigenous Australians, such as racism or other forms of discrimination, social marginalization and dispossession, chronic stress and/or exposure to violence (Cunningham 2010).

In India, Gaur et al. (2006) found that the prevalence of asthma among the rural, urban city and urban-slum adult population of Delhi were 13.34%, 7.9% and 11.92%, respectively. The current prevalence of asthma was more pronounced if there was a history of family atopy, or where more vegetable debris, poor ventilation, air pollution and increased human density were present. There was also an association with increased smoking habits. However, no significant difference was noted in asthma prevalence between those living in urban and rural areas (Gaur et al. 2006).

More recently, an allergy report from Austria indicated that people with higher education levels, or those who are more highly qualified for jobs and who live in urban areas are more likely to be affected by allergies than are people from lower socioeconomic levels or rural areas (Dorner et al. 2007).

Access to healthcare services is a marker of primary care quality, because acute episodes of asthma are avoidable if they are managed and appropriately treated in the community. Gaps in access to medical services between urban and rural areas exist, and include such things as convenience of transportation, range of services provided locally, as well as the cost for medical treatment. The previous literature indicates that a lack of medical services and specialists are more common in rural than in urban areas (Rural Healthy People 2010), and there is a low utilization efficiency of hospice services in rural areas (Gessert et al. 2006); in addition, disparities exist in the threshold for admission to hospital or clinic care, between urban and rural physicians (Russo et al. 1999).

Several studies worldwide have demonstrated the differences that exist in access to healthcare services for those suffering asthma morbidity. In Hawaii, 4,318 adults, whose names were selected from insurance claims data, were studied and the results showed that the rate of office visits after emergency department (ED) treatment for asthma was significantly lower in rural compared to urban residents. Rural residents were less likely to receive follow-up care after ED visits (RR: 78, 95% CI, 0.63–0.96, $p=0.02$; adjusted for age, sex, and morbidity). Distance to care centers and number of available healthcare providers, rather than financial aspects accounted for the differences (Withy and Davis 2008). Ansari et al. (2003) reported trends and geographic variations in hospital admissions for asthma sufferers in Victoria. Asthma admission rates were consistently higher in rural than in metropolitan areas. Rural doctors were more likely to admit people to hospital care after light attacks, because of concern that their conditions might deteriorate before they could return to the hospital (i.e., considering that they lived further from the hospital than city dwellers do; Ansari et al. 2003).

A cross-sectional study in Wellington, New Zealand was performed in which the regional variations in asthma hospital admission among Maori and non-Maori (Ellison-Loschmann et al. 2004) peoples were investigated. Asthma hospital admissions were more prevalent for Maoris than for non-Maoris of different age groups: 15–34 years relative risk (RR) = 1.31; 35–74 years RR = 2.97. The differences were somewhat higher in rural (RR: 1.34, 3.13 respectively) than in urban areas (RR: 1.22, 2.79 respectively). The disparities among the urban and rural groups in asthma morbidity for those living in rural areas included inadequate access to appropriate health care (e.g., inadequate money for traveling to the doctor's surgery, for paying doctor's fees, for buying drugs, etc.), and poor asthma education and differential management of asthma. Mugusi et al. (2004) performed another study in Tanzania and in Cameroon on the prevalence of wheezing and on self-reported asthma and asthma care in urban vs. rural areas of the two countries. There were no consistent patterns of urban: rural prevalence. Although consistent patterns of urban–rural prevalence were not obvious, peak flow rates varied with age, peaking at 25–34 years, and were higher in urban areas (age adjusted difference 22–70 L/min), and in the Tanzania populations. Asthma awareness (83–86% vs. 52–58%) and treatment (43–71% vs. 30–44%) of asthma was higher among those with a current wheeze in rural areas, compared with those in urban areas. Diagnosis by traditional healers and use of traditional remedies prevailed among self-reported asthmatic patients in rural Cameroon, and a major gap in clinical care, particularly in urban areas, was revealed (Mugusi et al. 2004).

Moreover, Roy found that asthma hospitalization rates were significantly higher in rural Delta regions of Mississippi compared with the urban Jackson Metropolitan Statistical Area, especially among older adults (≥65 years) and blacks. Blacks with asthma are more likely to have multiple asthma hospitalizations in Mississippi. Higher odds of multiple asthma discharges for Delta residents were not explained by race, sex, age, or income, indicating that other contributing factors (e.g., environmental, social, and access to care factors) need to be further investigated (Roy et al. 2010).

3 Conclusions

Asthma is a complex inflammatory disease of multifactorial origin, and is influenced by both environmental and genetic factors. The disparities in asthma prevalence and morbidity among the world's geographic locations are more likely to be associated with environmental exposures than genetic differences. In writing this article, we did find some studies that addressed the question of disparities in the prevalence of asthma morbidity in various geographic locations (see Table 1). Some studies addressed differences in prevalence of asthma morbidity in relation to various environmental or social factors (see Table 2). Most authors, who published on the topic of this paper, used a cross-sectional design in their experiments.

Table 1 Urban–rural differences in the prevalence of asthma

Study year	Reference and study location	Prevalence (%)	
		Urban area	Rural area
1994–1995	Filipiak et al. (2001), Germany	2.3	1.4
1994–2000	Ellison-Loschmann et al. (2004), New Zealand	1.4	1.7
1996	Yemaneberhan et al. (1997), Ethiopia	3.6	1.3
1996–1997	Walraven et al. (2001), Gambia	3.6	0.7
1999	Woods et al. (2000), Australia	9.0	12.0
1999	Venners et al. (2001), China	1.5	1.7
1999–2000	Viinanen et al. (2005), Mongolia	2.1	1.1
2000	Dejsomritrutai et al. (2006), Thailand	1.7	2.1
2000–2001	Ghosh et al. (2008), Canada	7.7	6.7
2001–2002	Iversen et al. (2005), Scotland	23.8	14.0
2005	Gaur et al. (2006), Delhi, India	10.2	13.3
2005	Morrison et al. (2009), the USA	7.8	7.9
2006–2007	Gao et al. (2011), Qinghai, China	0.3	0.6
2007–2008	Lâm et al. (2011), Vietnam	5.6	3.9
2008	Frazier et al. (2012), Montana, the USA	9.9	9.3
2008–2009	Musafiri et al. (2011), Rwanda	9.3	8.3

We reviewed 145 papers that pertained to the topic of this chapter. The key question we addressed is whether there is a difference in the prevalence of asthma or asthma-related symptoms for people who live in rural vs. urban areas. The main conclusions we arrived at on reviewing this question are as follows:

1. Indoor environments influence asthma and asthma-related symptoms among adults in both urban and rural areas.

   In most studies, reported IAQ and asthma morbidity data strongly indicated positive associations between indoor air pollution and adverse health effects in both urban and rural areas. Indoor factors most consistently associated with asthma and asthma-related symptoms in adults included fuel combustion, mold growth, and environmental tobacco smoke, in both rural and urban areas. Environmental exposures may increase an adult's risk of developing asthma and also may increase the risk of asthma exacerbations.

2. People living in urban areas often suffer greater asthma morbidity.

   Asthma and asthma-related symptoms occurred more frequently in urban than in rural areas, and that difference correlated with environmental risk exposures, SES, and healthcare access. Environmental risk factors to which urban adults were more frequently exposed than rural adults were dust mites, high levels of vehicle emissions, and a westernized lifestyle.

   Exposure to indoor biological contaminants in urban environment is common. A broad review of the literature for indoor environment generally suggests a relationship between microbe exposure and asthma prevalence and morbidity. The main risk factors for developing asthma in urban areas are atopy and allergy to house dust mites, followed by allergens from animal dander. House dust mite exposure may be a main reason for a higher diagnosis rate of asthma among adults living in urban vs. rural areas.

Table 2 A summary review of studies that have addressed factors that potentially affect asthma among adults residing in urban vs. rural areas

Potential factors	Reference, study year, and location	Study design, subject characteristics, and sample size (rural/urban)	Health outcomes	Risk assessment	Summary of published findings
Exposure to smoking	Ghosh et al. (2009), Canada	Cross-sectional study Residents aged 18–64 years ($n = 1,362/4,240$)	Asthma	Rural female smokers 1.4 times (95% CI, 1.02–1.94) more likely to be diagnosed with asthma	Living in a rural area and smoking increased asthma prevalence.
	Ghosh et al. (2008), Canada	Cross-sectional study Residents aged ≥ 15 years, 19,600 households ($n = 20,095$)	Asthma	–	Asthma prevalence increased among rural and urban women.
	Woods et al. (2000), Australia	Cross-sectional study Young adults 20–44 years ($n = 3,106/3,194$)	Nocturnal dyspnoea, asthma attack	–	Asthma prevalence was higher in rural areas.
	Walraven et al. (2001), Gambia	Cross-sectional study Residents aged ≥ 15 years ($n = 3,223/2,166$)	Asthma and chronic cough	–	The risk of asthma was not elevated in active smokers.
	Lâm et al. (2011) Bavi/Hanoi, Vietnam	Cross-sectional study Residents aged 21–70 years ($n = 4,000/3,008$)	Asthma	–	Asthma prevalence increased in both urban and rural areas.
Exposure to coal and biomass fuels	Venners et al. (2001), China	Cross-sectional study Residents aged ≥ 18 years ($n = 1,245/1,184$)	Respiratory symptoms (e.g., chronic cough)	–	Differences were found between study areas in the prevalence of respiratory symptoms.
	Shrestha and Shrestha (2005), Nepal	Cross-sectional study Residents, aged ≥ 18 years, 98 households ($n = 168$)	Cough, phlegm, dyspnoea and asthma	–	Prevalence of respiratory illnesses was higher in hills and rural areas.

(continued)

Table 2 (continued)

Potential factors	Reference, study year, and location	Study design, subject characteristics, and sample size (rural/urban)	Health outcomes	Risk assessment	Summary of published findings
Exposure to biological contaminants	Filipiak et al. (2001), German	Cross-sectional study Residents aged 25–75 years (n=1,470/3,110)	Allergic diseases	The urban population had an increased risk of allergic rhinitis (OR. 1.5; 95% CI, 1.2±1.9).	A farming environment might be protective for preventing allergies.
	Kurt et al. (2009), Turkey.	Cross-sectional study Parents of school children (n=10,289/15,554)	Asthma, wheezing, allergic rhinitis, and eczema	–	Living in a house with mold increased risk of respiratory illnesses.
	Linneberg (2005), Denmark	Cohort study Residents aged 15–69 years (n=231 (emigrant)/ 171(city resident))	Allergy	Migration from rural to urban areas increases risk of allergy (OR: 3.4, 95% CI, 1.6–7.2)	Migration from rural to urban areas increases risk of allergy.
	Bråbäck et al. (2004), Sweden	Cohort study Conscripts aged 17–20 years (n=197,547/1,119,437)	Asthma, allergic rhinitis, and eczema	Risk ratio for asthma in conscripts from farming vs. non-farming families was 1.00 (95% CI, 0.93–1.07)	Environmental changes increase in asthma in both farming and non-farming areas.
	Raukas-Kivioja et al. (2007), Estonia	Cross-sectional study Residents aged 17–69 years (n=651/541)	Allergic sensitization	Living in capital was associated with cat (OR:1.96, 95% CI, 1.03–3.74)	Urban living was associated with higher prevalence of allergic sensitization.
	Viinanen et al. (2005), Mongolian	Cross-sectional study Residents aged 10–60 years (n=304/896)	Asthma and allergic rhino conjunctivitis	–	Prevalence of atopic diseases was low in rural and increased with urbanization.
	Iversen et al. (2005), Scotland	Cross-sectional study Residents, aged ≥ 16 years (n=1,099/1,497)	Self-reported asthma and asthma morbidity	Residents from rural areas have lower asthma prevalence (OR: 0.59; 95% CI, 0.46–0.76)	Living in rural area gave lower prevalence of asthma and wheeze.
	Kilpeläinen et al. (2001), Finland	Cross-sectional study First-year university students aged 18–25 years (n=651/541)	Asthma and sensitization	Current asthma had a risk ratio of 0.22 (95% CI, 0.07–0.70) in subjects with childhood farm environment	Farm environment in childhood protects against adult asthma.
	Lourenço et al. (2009), Portugal	Cross-sectional study Patients aged ≥ 18 years (n=268/226)	Asthma and allergic rhinitis	–	Allergic rhinitis prevalence was higher in urban asthmatic patients.

Category	Reference, country	Study design, sample	Outcome	Result	Conclusion
Air pollution, geographical and meteorological conditions	Cingi et al. (2010), Turkey	Cross-sectional study Caucasian inhabitants, aged $\geq$ 18 years ($n=499/4,125$)	Allergic rhinitis	—	AR prevalence in coastal urban areas was higher than that in inner rural areas.
	Dejsomritrutai et al. (2006), Thailand	Cross-sectional study Residents, aged 20–44 years ($n=3,452$)	Bronchial hyper responsiveness and asthma	—	Asthma prevalence was lower in southeast Asian countries.
	de Marco et al. (2002), Italy	Cross-sectional study Residents aged 20–44 years ($n=18,873$)	Asthma and allergic rhinitis	Mediterranean climate increased asthma attack risk (OR: 1.19; 95% CI 1.07–1.31)	Traffic air pollution might pose long-term adverse effects on asthma morbidity.
Socioeconomic characteristics and healthcare services	Mugusi et al. (2004), Tanzania and Cameroon	Cross-sectional study Residents, aged $\geq$ 16 years ($n=4,560$)	Wheeze and self-reported asthma	—	Asthma was an important clinical condition in sub-Saharan Africa.
	Withy and Davis, (2008), the USA	Cross-sectional study Residents, aged $\geq$ 18 years ($n=2,652/4,412$)	Asthma	Rural residents were less likely to receive follow-up care (RR: 78, 95% CI, 0.63–0.96)	Rural residents were less likely to receive follow-up care for a diagnosis of asthma.
	Ansari et al. (2003), Australia	Cross-sectional study Residents ($n=10,079$)	Asthma	—	Rural areas had higher asthma hospital admission rates.
	Ellison-Loschmann et al. (2004), New Zealand	Cross-sectional study Maori and non-Maori ($n=6,036/19,829$)	Asthma	Asthma hospitalization rates were higher in rural (RR: 3.13) than in urban areas (RR: 2.79)	Asthma hospitalization rates were higher in rural than in urban areas.
	Morrison et al. (2009), the USA	Cross-sectional study Residents, aged $\geq$ 18 years ($n=341,932$)	Asthma	Asthma prevalence (7.9%; 95% CI: 7.73–8.08) was not statistically different	Asthma prevalence was as high in rural as in urban areas.
	Cunningham (2010), Australia	Cross-sectional study Indigenous and nonindigenous adults, aged 18–64 years ($n=5,417/15,432$)	Asthma	—	The socioeconomic patterning of asthma among indigenous Australians is lower.
	Gaur et al. (2006), India	Cross-sectional study Residents, aged 18–70 years ($n=1,552/1,876$)	Asthma and allergic rhinitis	—	No difference in asthma prevalence between urban and rural population.

In addition, the prevalence of asthma morbidity increases with urbanization. High levels of vehicle emissions, Western lifestyles and degree of urbanization itself, may affect outdoor, and thereby indoor air quality. In urban areas, biomass fuels have been widely replaced by cleaner energy sources at home, such as gas and electricity, but in most developing countries, coal is still a major source of fuel for cooking and heating, particularly in winter. Noticeably, chemical emissions from new building materials and furniture, such as formaldehyde, are problematic in urban settings. Moreover, exposure to ETS is common at home or at work in urban areas.

3. The prevalence of asthma morbidity is less common in rural areas.

There is evidence that asthma prevalence and morbidity is less common in rural than in urban areas. The possible reasons are that rural residents are exposed early in life to stables and to farm milk production, and such exposures are protective against developing asthma morbidity. Even so, asthma morbidity is disproportionately high among poor inner-city residents and in rural populations. A higher proportion of adult residents of nonmetropolitan areas were characterized as follows: aged 55 years or older, no previous college admission, low household income, no health insurance coverage, and could not see a doctor due to healthcare service availability, etc.

In rural areas, biomass fuels meet more than 70% of the rural energy needs (China Statistical Yearbook 1998). Progress in adopting modern energy sources in rural areas has been slow. Household use of traditional biomass fuels including firewood, wood chips, crop residue, dung cakes and coal are still widespread in rural areas. The rural poor depend upon biomass fuels for their basic cooking, and water- and space-heating needs. The most direct health impact comes from household energy use among the poor, who depend almost entirely on burning biomass fuels in simple cooking devices that are placed in inadequately ventilated spaces. Future research is needed to assess the long-term effects of biomass smoke on lung health among adults in rural areas. Given the extensive use of biomass fuels in rural areas, public health efforts in the developing world that are concerned with respiratory health should address the risks IAP exposure. In Fig. 1, we summarize the underlying susceptibility factors that contribute to discrepancies in the prevalence of asthma morbidity among adults in urban vs. rural areas.

Geographic differences in asthma susceptibility exist around the world. The reason for the differences in asthma prevalence in rural and urban areas may be due to the fact that populations have different lifestyles and cultures, as well as different environmental exposures and different genetic backgrounds. Identifying geographic disparities in asthma hospitalizations is critical to implementing prevention strategies, reducing morbidity, and improving healthcare financing for clinical asthma treatment. Although evidence shows that differences in the prevalence of asthma do exist between urban and rural dwellers in many parts of the world, including in developed countries, data are inadequate to evaluate the extent to which different pollutant exposures contribute to asthma morbidity and severity of asthma between urban and rural areas.

Finally, we addressed potential biases of the studies reviewed to prepare this paper. Most reviewed studies utilized a cross-sectional design, which limits their use for inferring causality of effect. What constituted an urban or rural area and how these were defined in terms of population density or numbers varied considerable among reviewed studies. Therefore, it is difficult to compare studies conducted in widely different geographies. Furthermore, comparisons are different because populations have different lifestyles and cultures, as well as different environmental exposures and different genetic backgrounds. We suspect that the differences in how urban or rural areas were defined in different studies have affected some findings of the association between increased respiratory symptoms from indoor exposures or risk factors. In previous studies, synergism or additive effects from multiple indoor environmental exposures have been reported. In contrast, few study authors of papers in this review claimed to have observed interactive effects from multifactor exposures in either rural or urban surveys.

4   Summary

In this review, our aim was to examine the influence of geographic variations on asthma prevalence and morbidity among adults, which is important for improving our understanding, identifying the burden, and for developing and implementing interventions aimed at reducing asthma morbidity. Asthma is a complex inflammatory disease of multifactorial origin, and is influenced by both environmental and genetic factors. The disparities in asthma prevalence and morbidity among the world's geographic locations are more likely to be associated with environmental exposures than genetic differences. In writing this article, we found that the indoor factors most consistently associated with asthma and asthma-related symptoms in adults included fuel combustion, mold growth, and environmental tobacco smoke in both urban and rural areas. Asthma and asthma-related symptoms occurred more frequently in urban than in rural areas, and that difference correlated with environmental risk exposures, SES, and healthcare access. Environmental risk factors to which urban adults were more frequently exposed than rural adults were dust mites, high levels of vehicle emissions, and a westernized lifestyle.

Exposure to indoor biological contaminants in the urban environment is common. The main risk factors for developing asthma in urban areas are atopy and allergy to house dust mites, followed by allergens from animal dander. House dust mite exposure may potentially explain differences in diagnosis of asthma prevalence and morbidity among adults in urban vs. rural areas. In addition, the prevalence of asthma morbidity increases with urbanization. High levels of vehicle emissions, Western lifestyles and degree of urbanization itself, may affect outdoor and thereby indoor air quality. In urban areas, biomass fuels have been widely replaced by cleaner energy sources at home, such as gas and electricity, but in most developing countries, coal is still a major source of fuel for cooking and heating, particularly in winter. Moreover, exposure to ETS is common at home or at work in urban areas.

There is evidence that asthma prevalence and morbidity is less common in rural than in urban areas. The possible reasons are that rural residents are exposed early in life to stables and to farm milk production, and such exposures are protective against developing asthma morbidity. Even so, asthma morbidity is disproportionately high among poor inner-city residents and in rural populations. A higher proportion of adult residents of nonmetropolitan areas were characterized as follows: aged 55 years or older, no previous college admission, low household income, no health insurance coverage, and could not see a doctor due to healthcare service availability, etc. In rural areas, biomass fuels meet more than 70% of the rural energy needs. Progress in adopting modern energy sources in rural areas has been slow. The most direct health impact comes from household energy use among the poor, who depend almost entirely on burning biomass fuels in simple cooking devices that are placed in inadequately ventilated spaces. Prospective studies are needed to assess the long-term effects of biomass smoke on lung health among adults in rural areas.

Geographic differences in asthma susceptibility exist around the world. The reason for the differences in asthma prevalence in rural and urban areas may be due to the fact that populations have different lifestyles and cultures, as well as different environmental exposures and different genetic backgrounds. Identifying geographic disparities in asthma hospitalizations is critical to implementing prevention strategies, reducing morbidity, and improving healthcare financing for clinical asthma treatment. Although evidence shows that differences in the prevalence of asthma do exist between urban and rural dwellers in many parts of the world, including in developed countries, data are inadequate to evaluate the extent to which different pollutant exposures contribute to asthma morbidity and severity of asthma between urban and rural areas.

Acknowledgments This work was supported by the foundation of: National University of Malaysia, Kuala Lumpur, Malaysia (FF-013-2012); Key Technologies R&D Programme of Department of Science and Technology of Guizhou Province, China (SY[2012]3126) and (SY[2011]3029).

References

Aggarwal AN, Chaudhry K, Chhabra SK, D'Souza GA, Gupta D, Jindal SK, Katiyar SK, Kumar R, Shah B, Vijayan VK; Asthma Epidemiology Study Group (2006) Prevalence and risk factors for bronchial asthma in Indian adults: a multicentre study. Indian J Chest Dis Allied Sci 48(1):13–22

Anastassakis KK, Chatzimichail A, Androulakis I, Charisoulis S, Riga M, Eleftheriadou A (2010) Skin prick test reactivity to common aeroallergens and ARIA classification of allergic rhinitis in patients of Central Greece. Eur Arch Otorhinolaryngol 267:77–85

Ansari Z, Haby MM, Henderson T, Cicuttini F, Ackland MJ (2003) Trends and geographic variations in hospital admissions for asthma in Victoria. Opportunities for targeted interventions. Aust Fam Physician 32:286–288

Arif AA, Borders TF, Patterson PJ, Rohrer JE, Xu KT (2004) Prevalence and correlates of paediatric asthma and wheezing in a largely rural USA population. J Paediatr Child Health 40:189–194

Armentia A, Lombardero M, Callejo A, Barber D, Martín Gil FJ, Martín-Santos JM, Vega JM, Arranz ML (2002) Is Lolium pollen from an urban environment more allergenic than rural pollen? Allergol Immunopathol (Madr) 30:218–224

ASHRAE (1989) ASHRAE Standard 62-1989, ventilation for acceptable indoor air quality. ASHRAE, Atlanta, Ga

Bosch-Cano F, Bernard N, Sudre B, Gillet F, Thibaudon M, Richard H, Badot PM, Ruffaldi P (2011) Human exposure to allergenic pollens: a comparison between urban and rural areas. Environ Res 111:619–625

Bousquet J, Khaltaev N, Cruz AA, Denburg J, Fokkens WJ, Togias A (2008) Allergic Rhinitis and its Impact on Asthma (ARIA) 2008 update (in collaboration with the World Health Organization, GA(2)LEN and AllerGen). Allergy 63:8–160

Bråbäck L, Hjern A, Rasmussen F (2004) Trends in asthma, allergic rhinitis and eczema among Swedish conscripts from farming and non-farming environments. A nationwide study over three decades. Clin Exp Allergy 34:38–43

Bruce N, Perez-Padilla R, Albalak R (2000) Indoor air pollution in developing countries: a major environmental and public health challenge. Bull World Health Organ 78:1078–1092

Busse WW, Lemanske RJ (2001) Asthma. N Engl J Med 344:350–362

Chen CH, Xirasagar S, Lin HC (2006) Seasonality in adult asthma admissions, air pollutant levels, and climate: a population-based study. J Asthma 43:287–292

China Statistical Yearbook 1997 (1998) State Statistical Bureau, People's Republic of China

Cingi C, Topuz B, Songu M, Kara CO, Ural A, Yaz A, Yildirim M, Miman MC, Bal C (2010) Prevalence of allergic rhinitis among the adult population in Turkey. Acta Otolaryngol 130(5):600–606.

Colbeck I, Nasir ZA, Ali Z, Ahmad S (2010a) Nitrogen dioxide and household fuel use in the Pakistan. Sci Total Environ 409:357–363

Colbeck I, Nasir ZA, Ali Z (2010b) Characteristics of indoor/outdoor particulate pollution in urban and rural residential environment of Pakistan. Indoor Air 20:40–51

Comhair SA, Gaston BM, Ricci KS, Hammel J, Dweik RA, Teague WG, Meyers D, Ampleford EJ, Bleecker ER, Busse WW, Calhoun WJ, Castro M, Chung KF, Curran-Everett D, Israel E, Jarjour WN, Moore W, Peters SP, Wenzel S, Hazen SL, Erzurum SC (2011) Detrimental effects of environmental tobacco smoke in relation to asthma severity. PLoS One 6:e18574

Cooper PJ, Rodrigues LC, Cruz AA, Barreto ML (2009) Asthma in Latin America: a public heath challenge and research opportunity. Allergy 64:5–17

Cunningham J (2010) Socioeconomic status and self-reported asthma in Indigenous and non-Indigenous Australian adults aged 18–64 years: analysis of national survey data. Int J Equity Health 9:18

de Marco R, Poli A, Ferrari M (2002) Italian study on asthma in young adults. The impact of climate and traffic-related NO_2 on the prevalence of asthma and allergic rhinitis in Italy. Clin Exp Allergy 32:1405–1412

Dejsomritrutai W, Nana A, Chierakul N, Tscheikuna J, Sompradeekul S, Ruttanaumpawan P, Charoenratanakul S (2006) Prevalence of bronchial hyperresponsiveness and asthma in the adult population in Thailand. Chest 129:602–609

Dorner T, Lawrence K, Rieder A, Kunze M (2007) Epidemiology of allergies in Austria. Results of the first Austrian allergy report. Wien Med Wochenschr 157:235–242

Dottorini ML, Bruni B, Peccini F, Bottini P, Pini L, Donato F, Casucci G, Tantucci C (2007) Skin prick-test reactivity to aeroallergens and allergic symptoms in an urban population of central Italy: a longitudinal study. Clin Exp Allergy 37:188–196

Douwes J, Travier N, Huang K, Cheng S, McKenzie J, Le Gros G, von Mutius E, Pearce N (2007) Lifelong farm exposure may strongly reduce the risk of asthma in adults. Allergy 62(10):1158–1165

Eduard W, Douwes J, Omenaas E, Heederik D (2004) Do farming exposures cause or prevent asthma? Results from a study of adult Norwegian farmers. Thorax 59:381–386

Ellison-Loschmann L, King R, Pearce N (2004) Regional variations in asthma hospitalisations among Maori and non-Maori. N Z Med J 117:U745

El-Shazly AM, El-Beshbishi SN, Azab MS, El-Nahas HA, Soliman ME, Fouad MA, Mel-S M (2006) Present situation of house dust mites in Dakahlia Governorate, Egypt. J Egypt Soc Parasitol 36:113–126

Filipiak B, Heinrich J, Schfer T, Ring J, Wichmann HE (2001) Farming, rural lifestyle and atopy in adults from southern Germany–results from the MONICA/KORA study Augsburg. Clin Exp Allergy Filipiak 31:1829–1838

Franco JM, Gurgel R, Sole D, Lúcia França V, Brabin B (2009) Socio-environmental conditions and geographical variability of asthma prevalence in Northeast Brazil. Allergol Immunopathol (Madr) 37:116–121

Frazier JC, Loveland KM, Zimmerman HJ, Helgerson SD, Harwell TS (2012) Prevalence of asthma among adults in metropolitan versus nonmetropolitan areas in Montana, 2008. Prev Chronic Dis 9:E09

Fukutomi Y, Nakamura H, Kobayashi F, Taniguchi M, Konno S, Nishimura M, Kawagishi Y, Watanabe J, Komase Y, Akamatsu Y, Okada C, Tanimoto Y, Takahashi K, Kimura T, Eboshida A, Hirota R, Ikei J, Odajima H, Nakagawa T, Akasawa A, Akiyama K (2010) Nationwide cross-sectional population-based study on the prevalences of asthma and asthma symptoms among Japanese adults. Int Arch Allergy Immunol 53:280–287

Fullerton DG, Semple S, Kalambo F, Suseno A, Malamba R, Henderson G, Ayres JG, Gordon SB (2009) Biomass fuel use and indoor air pollution in homes in Malawi. Occup Environ Med 66:777–783

Gao F, Yang QJ, Zhang HG (2011) An epidemiological study of bronchial asthma in Qinghai Province. Zhonghua Jie He He Hu Xi Za Zhi 34(3):165–168

Gaur K, Gupta S, Rajpal AB, Singh A, Rohatgi JN (2006) Prevalence of bronchial asthma and allergic rhinitis among urban and rural adult population of Delhi Indian Asthma. Indian J Allergy Asthma Immunol 20:90–97

Gessert CE, Haller IV, Kane RL, Degenholtz H (2006) Rural-urban differences in medical care for nursing home residents with severe dementia at the end of life. J Am Geriatr Soc 54:1199–1205

Ghosh S, Pahwa P, Rennie D, McDuffie HH (2008) Opposing trends in the prevalence of health professional-diagnosed asthma by sex: a Canadian National Population Health Survey study. Can Respir J 15:146–152

Ghosh S, Punam P, Donna CR, Bonnie J (2009) Gender-related interactive effect of smoking and rural/urban living on asthma prevalence: a longitudinal Canadian NPHS study. J Asthma 46:988–994

Guangzhou Statistics Yearbook 2002–2006 (2007) Guangzhou Municipal Statistics Bureau

Guinea J, Teresa P, Luis A, Emilio B (2006) Outdoor environmental levels of Aspergillus spp. conidia over a wide geographical area. Med Mycol 44:349–356

Hales S, Lewis S, Slater T, Crane J, Pearce N (1998) Prevalence of adult asthma symptoms in relation to climate in New Zealand. Environ Health Perspect 106:607–610

Haverkos HW (2004) Viruses, chemicals and co-carcinogenesis. Oncogene 23:6492–6499

Heinrich J, Wichmann HE (2004) Traffic related pollutants in Europe and their effect on allergic disease. Curr Opin Allergy Clin Immunol 4:341–348

Hersoug L, Lise LH, Torben S, Flemming M, Allan L (2010) Indoor exposure to environmental cigarette smoke, but not other inhaled particulates associates with respiratory symptoms and diminished lung function in adults. Respirology 15:993–1000

Ho MG, Ma S, Chai W, Xia W, Yang G, Novotny TE (2010) Smoking among rural and urban young women in China. Tob Control 19:13–18

Ige OM, Arinola OG, Oluwole O, Falade AG, Falusi AG, Aderemi T, Huo D, Olopade OI, Olopade CO (2011) Atopy is associated with asthma in adults living in rural and urban southwestern Nigeria. J Asthma 48(9):894–900

Iversen L, Hannaford PC, Price DB, Godden DJ (2005) Is living in a rural area good for your respiratory health? Results from a cross-sectional study in Scotland. Chest 128:2059–2067

Jackson JE, Doescher MP, Hart LG (2007) A national study of lifetime asthma prevalence and trends in metro and non-metro counties, 2000–2003. University of Washington, WWAMI Rural Health Research Center

Jarvis DL, Leaderer BP, Chinn S, Burney PG (2005) Indoor nitrous acid and respiratory symptoms and lung function in adults. Thorax 60(6):474–479

Jiang R, Bell ML (2008) A comparison of particulate matter from biomass-burning rural and non-biomass burning urban households in Northeastern China. Environ Health Perspect 116:907–914

Joshi K, Rajesh K, Ajit A (2003) Morbidity profile and its relationship with disability and psychological distress among elderly people in Northern India. Int J Epidemiol 32:978–987

Kasprzyk I, Worek M (2006) Airborne fungal spores in urban and rural environments in Poland. Aerobiologia 22:169–176

Kilpeläinen M, Terho EO, Helenius H, Koskenvuo M (2001) Home dampness, current allergic diseases, and respiratory infections among young adults. Thorax 56:462–467

Krawczyk P, Chocholska S, Mackiewicz B, Trembas-Pietraś E, Wegrzyn-Szkutnik I, Milanowski J (2003) Differences in clinical test results of patients with bronchial asthma and living environment. Research in rural and urban areas in the eastern part of Poland. Ann Univ Mariae Curie Sklodowska Med 58:452–458

Kurt E, Metintas S, Basyigit I, Bulut I, Coskun E, Dabak S (2009) Prevalence and risk factors of allergies in Turkey (PARFAIT): results of a multicentre cross-sectional study in adults. Eur Respir J 33:724–733

Lâm HT, Rönmark E, Tu'ò'ng NV, Ekerljung L, Chúc NT, Lundbäck B (2011) Increase in asthma and a high prevalence of bronchitis: results from a population study among adults in urban and rural Vietnam. Respir Med 105:177–185

Lei YC, Chan CC, Wang PY, Lee CT, Cheng TJ (2004) Effects of Asian dust event particles on inflammation markers in peripheral blood and bronchoalveolar lavage in pulmonary hypertensive rats. Environ Res 95:71–76

Lewis S, Hales S, Slater T, Pearce N, Crane J, Beasley R (1997) Geographical variation in the prevalence of asthma symptoms in New Zealand. N Z Med J J110(1049):286–289

Linneberg A (2005) Hypothesis: urbanization and the allergy epidemic–a reverse case of immunotherapy? Allergy 60:538–539

Liu AH (2004) Something old, something new: indoor endotoxin, allergens and asthma. Paediatr Respir Rev 5 Suppl A:S65–71

Loureiro G, Rabaça MA, Blanco B, Andrade S, Chieira C, Pereira C (2005) Urban versus rural environment–any differences in aeroallergens sensitization in an allergic population of Cova da Beira, Portugal? Eur Ann Allergy Clin Immunol 37(5):187–193

Lourenço O, Fonseca AM, Taborda-Barata L (2009) Asthma is more frequently associated with non-allergic than allergic rhinitis in Portuguese patients. Rhinology 47:207–213

Lu N, Samuels ME, Kletke PR, Whitler ET (2010) Rural-urban differences in health insurance coverage and patterns among working-age adults in Kentucky. J Rural Health 26(2):129–138

Matson U (2005) Indoor and outdoor concentrations of ultrafine particles in some Scandinavian rural and urban areas. Sci Total Environ 343:169–176

Mestl HE, Aunan K, Seip HM, Wang SX, Zhao Y, Zhang D (2007a) Urban and rural exposure to indoor air pollution from domestic biomass and coal burning across China. Sci Tot Environ 377:12–26

Mestl HE, Aunan K, Seip HM (2007b) Health benefits from reducing indoor air pollution from household solid fuel use in China—three abatement scenarios. Environ Int 33:831–840

Metintas S, Kurt E (2010) Geo-climate effects on asthma and allergic diseases in adults in Turkey: results of PARFAIT study. Int J Environ Health Res 20:189–199

Michnar M, Trembas-Pietraś E, Milanowski J (2003) Incidence of allergic respiratory diseases in the Lublin region in correlation with the level of environmental pollution. Ann Univ Mariae Curie Sklodowska Med 58(1):481–487

Mishra V (2003) Effect of indoor air pollution from biomass combustion on prevalence of asthma in the elderly. Environ Health Perspect 111:71–78

Modig L, Järvholm B, Rönnmark E, Nyström L, Lundbäck B, Andersson C, Forsberg B (2006) Vehicle exhaust exposure in an incident case-control study of adult asthma. Eur Respir J 28(1):75–81

Morrison T, Callahan D, Moorman J, Bailey C (2009) A national survey of adult asthma prevalence by urban-rural residence U.S. 2005. J Asthma 46:751–758

Mugusi F, Edwards R, Hayes L, Unwin N, Mbanya JC, Whiting D, Sobngwi E, Rashid S (2004) Prevalence of wheeze and self-reported asthma and asthma care in an urban and rural area of Tanzania and Cameroon. Trop Doct 34:209–214

Musafiri S, van Meerbeeck J, Musango L, Brusselle G, Joos G, Seminega B (2011) Prevalence of atopy, asthma and COPD in an urban and a rural area of an African country. Respir Med 105:1596–1605

Nguyen T, Lurie M, Gomez M, Reddy A, Pandya K, Medvesky M (2010) The National Asthma Survey–New York State: association of the home environment with current asthma status. Public Health Rep 125:877–887

Nguyen K, Zahran H, Iqbal S, Peng J, Boulay E (2011) Factors associated with asthma control among adults in five New England states, 2006–2007. J Asthma 48:581–588

Nyan OA, Walraven GE, Banya WA, Milligan P, Van Der Sande M, Ceesay SM (2001) Atopy, intestinal helminth infection and total serum IgE in rural and urban adult Gambian communities. Clin Exp Allergy 31:1672–1678

Oliveira AC, Martins AN, Pires AF, Arruda MB, Tanuri A, Pereira HS, Brindeiro RM (2009) Enfuvirtide (T-20) resistance-related mutations in HIV type 1 subtypes B, C, and F isolates from Brazilian patients failing HAART. AIDS Res Hum Retroviruses 25:193–198

Persinger RL, Blay WM, Heintz NH, Hemenway DR, Janssen-Heininger YM (2001) Nitrogen dioxide induces death in lung epithelial cells in a density-dependent manner. Am J Respir Cell Mol Biol 24(5):583–590

Peterson JS (2012) Nitrous dioxide toxicity—medscape. http://emedicine.medscape.com/article/820431-overview. Accessed 7 Aug 2012

Polosa R, Russo C, Caponnetto P, Bertino G, Sarvà M, Antic T, Mancuso S, Al-Delaimy WK (2011) Greater severity of new onset asthma in allergic subjects who smoke: a 10-year longitudinal study. Respir Res 12:16

Qian Z, He Q, Kong L, Xu F, Wei F, Chapman RS, Chen W, Edwards RD, Bascom R (2007) Respiratory responses to diverse indoor combustion air pollution sources. Indoor Air 17:135–142

Raukas-Kivioja A, Raukas ES, Meren M, Loit HM, Rönmark E, Lundbäck B (2007) Allergic sensitization to common airborne allergens among adults in Estonia. Int Arch Allergy Immunol 142:247–254

Riva DR, Magalhães CB, Lopes AA, Lanças T, Mauad T, Malm O, Valença SS, Saldiva PH, Faffe DS, Zin WA (2011) Low dose of fine particulate matter (PM2.5) can induce acute oxidative stress, inflammation and pulmonary impairment in healthy mice. Inhal Toxicol 23:257–267

Romans S, Cohen M, Forte T (2011) Rates of depression and anxiety in urban and rural Canada. Soc Psychiatry Psychiatr Epidemiol 46(7):567–575

Roy SR, Schiltz AM, Marotta A, Shen Y, Liu AH (2003) Bacterial DNA in house and farm barn dust. J Allergy Clin Immunol 112(3):571–578

Roy SR, McGinty EE, Hayes SC, Zhang L (2010) Regional and racial disparities in asthma hospitalizations in Mississippi. J Allergy Clin Immunol 125(3):636–642

Rural Healthy People (2010) Healthy people 2010: a companion document for rural areas. http://www.studenthere.com/redirect.php? Accessed 25 June 2011

Russo MJ, McConnochie KM, McBride JT, Szilagyi PG, Brooks AM, Roghmann KJ (1999) Increase in admission threshold explains stable asthma hospitalization rates. Pediatrics 104:454–462

Sandstrom T, Helleday R, Bjermer L, Stjernberg N (1992) Effects of repeated exposure to 4 ppm nitrogen dioxide on bronchoalveolar lymphocyte subsets and macrophages in healthy men. Eur Respir J 5:1092–1096

Senthilselvan A, Lawson J, Rennie DC, Dosman JA (2003) Stabilization of an increasing trend in physician-diagnosed asthma prevalence in Saskatchewan, 1991 to 1998. Chest 124:438–448

Shrestha IL, Shrestha SL (2005) Indoor air pollution from biomass fuels and respiratory health of the exposed population in Nepalese households. Int J Occup Environ Health 11:150–160

Skorge TD, Eagan TM, Eide GE, Gulsvik A, Bakke PS (2005) Indoor exposures and respiratory symptoms in a Norwegian community sample. Thorax 60:937–942

Smith K, Warholak T, Armstrong E, Leib M, Rehfeld R, Malone D (2009) Evaluation of risk factors and health outcomes among persons with asthma. J Asthma 46:234–237

Solomon C, Christian DL, Chen LL, Welch BS, Kleinman MT, Dunham E, Erle DJ, Balmes JR (2000) Effect of serial-day exposure to nitrogen dioxide on airway and blood leukocytes and lymphocyte subsets. Eur Respir J 15:922–928

Spengler JD, Ferris BG, Dockery DW, Speizer FE (1979) Sulfur dioxide and nitrogen dioxide levels inside and outside homes and the implications on health effects research. Environ Sci Technol 13:1276–1280

Taksey J, Craig TJ (2001) Allergy test results of a rural and small-city population compared with those of an urban population. J Am Osteopath Assoc 101:S4–S7

Trupin L, Balmes JR, Chen H, Eisner MD, Hammond SK, Katz PP, Lurmann F, Quinlan PJ, Thorne PS, Yelin EH, Blanc PD (2010) An integrated model of environmental factors in adult asthma lung function and disease severity: a cross-sectional study. Environ Health 9:24

United States Environmental Protection Agency (USEPA) (1993) Fact sheet: respiratory health effects of passive smoking. http://www.epa.gov/smokefree/pubs/etsfs.html. Accessed 22 June 2011

Valet RS, Perry TT, Hartert TV (2009) Rural health disparities in asthma care and outcomes. J Allergy Clin Immunol 123(6):1220–1225

Vander Weg MW, Cunningham CL, Howren MB, Cai X (2011) Tobacco use and exposure in rural areas: findings from the behavioral risk factor surveillance system. Addict Behav 36:231–236

Venners SA, Wang B, Ni J, Jin Y, Yang J, Fang Z, Xu X (2001) Indoor air pollution and respiratory health in urban and rural China. Int J Occup Environ Health 7:173–181

Verlato G, Calabrese R, De Marco R (2002) Correlation between asthma and climate in the European Community Respiratory Health Survey. Arch Environ Health 57:48–52

Viinanen A, Munhbayarlah S, Zevgee T, Narantsetseg L, Naidansuren T, Koskenvuo M, Helenius H, Terho EO (2005) Prevalence of asthma, allergic rhinoconjunctivitis and allergic sensitization in Mongolia. Allergy 60:1370–1377

Viinanen A, Munhbayarlah S, Zevgee T, Narantsetseg L, Naidansuren T, Koskenvuo M, Helenius H, Terho EO (2007) The protective effect of rural living against atopy in Mongolia. Allergy 62:272–280

von Mutius E, Vercelli D (2010) Farm living: effects on childhood asthma and allergy. Nat Rev Immunol 10:861–868

von Mutius E, Braun-Fahrländer C, Schierl R, Riedler J, Ehlermann S, Maisch S, Waser M, Nowak D (2000) Exposure to endotoxin or other bacterial components might protect against the development of atopy. Clin Exp Allergy 30:1230–1234

Walraven GE, Nyan OA, Van Der Sande MA, Banya WA, Ceesay SM, Milligan PJ, McAdam KP (2001) Asthma, smoking and chronic cough in rural and urban adult communities in The Gambia. Clin Exp Allergy 31:1679–1685

Withy K, Davis J (2008) Followup after an emergency department visit for asthma: urban/rural patterns. Ethn Dis 18:S2–S247

Woods RK, Burton DL, Wharton C, McKenzie GH, Walters EH, Comino EJ, Abramson MJ (2000) Asthma is more prevalent in rural New South Wales than metropolitan Victoria, Australia. Respirology 5:257–263

Xiao L, Yang Y, Li Q, Wang CX, Yang GH (2010) Population-based survey of secondhand smoke exposure in China. Biomed Environ Sci 23:430–436

Yemaneberhan H, Bekele Z, Venn A, Lewis S, Parry E, Britton J (1997) Prevalence of wheeze and asthma and relation to atopy in urban and rural Ethiopia. Lancet 350:85–90

Yu J, Huang HJ, Jin F, Xu J (2011a) The role of airborne microbes in school and its impact on asthma, allergy and respiratory symptoms among school children. Rev Med Microbiol 22:84–89

Yu J, Noor HI, Xu J, Zaleha MI (2011b) Do indoor environments influence on asthma and asthma-related symptoms among adults in homes? A review of the literature. J Formos Med Assoc 110:555–563

Zhang LX, Enarson DA, He GX, Li B, Chan-Yeung M (2002) Occupational and environmental risk factors for respiratory symptoms in rural Beijing, China. Eur Respir J 20:1525–1531

Mercury in the Atmospheric and Coastal Environments of Mexico

Jorge Ruelas-Inzunza, Carolina Delgado-Alvarez,
Martín Frías-Espericueta, and Federico Páez-Osuna

Contents

1 Introduction

Though mercury (Hg) occurs naturally in the environment, anthropogenic activities have affected its global cycle in ways that mobilize increasing amounts of this metal; currently, such human-related activities mobilize more Hg than do natural

J. Ruelas-Inzunza (✉)
Technological Institute of Mazatlán, P.O. Box 757, Mazatlán, Sinaloa 82000, Mexico
e-mail: ruelas@ola.icmyl.unam.mx

C. Delgado-Alvarez • M. Frías-Espericueta
Faculty of Marine Sciences, Autonomous University of Sinaloa, Mazatlán, Sinaloa 82000, Mexico

F. Páez-Osuna
Instituto de Ciencias del Mar y Limnología, Universidad Nacional Autónoma de México, P.O. Box 811, Mazatlán, Sinaloa 82000, Mexico

D.M. Whitacre (ed.), *Reviews of Environmental Contamination and Toxicology*
Volume 226, Reviews of Environmental Contamination and Toxicology 226,
DOI 10.1007/978-1-4614-6898-1_3, © Springer Science+Business Media New York 2013

processes (Fitzgerald and Lamborg 2005). It has been estimated that the quantity of Hg mobilized into the atmosphere has increased from two to five times (Nriagu and Pacyna 1988) since the beginning of the industrial age. The mercury cycle is complex and involves diverse environmental media that include air, land, and water. For any country that is not landlocked, the estuaries and coastal waters constitute an important link between the terrestrial environment and the open oceanic waters (Mason et al. 1994). However, research thus far performed on Hg as an environmental contaminant has been focused mainly on terrestrial ecosystems (Fitzgerald and Mason 1996). The focus on land contamination by Hg has occurred despite the prominent role played by Hg in the atmosphere (transported by wind and deposited under both dry and wet conditions) and in oceanic processes (horizontal and vertical transportation, accumulation in sediments, and bacterial transformations).

Within the framework of the North America Free Trade Agreement (NAFTA), established among Canada, the United States of America, and Mexico, the Environmental Cooperation Commission (ECC) was created to contribute to solving environmental problems. A relevant issue concerning the environment and human health is the reduction of Hg emissions. Such emissions result from different anthropogenic activities that include industrial and commercial processes, and urban activities. It has been estimated that from 1540 to 1850, about 45,000 t of Hg were sent from Spain to Mexico for use in extraction of gold and silver from mines in Zacatecas and other cities of central Mexico (De-la-Peña-Sobarzo 2003). Although significant uncertainty remains about the occurrence and levels of Hg in the environment, it appears that the global movement of this metal primarily involves inorganic forms (WHO 1990).

The presence of environmental residues of Hg is an issue of concern in many countries; in Mexico, studies that relate to the occurrence of this element are limited. In this review, we address published information on Hg as a pollutant and its presence in diverse environmental compartments. Our aim is to first review the natural and anthropogenic sources of Hg pollution in Mexico. Then, we address the levels of Hg that appear in the atmospheric and aquatic environments of Mexico. Finally, we address how Hg interacts with biota, including invertebrates, vertebrates, and other taxonomic groups.

2   Sources of Mercury

There are four main environmental sources of Hg (PNUMA 2005): (1) natural, (2) anthropogenic releases from mobilizing Hg impurities that exist in raw materials (e.g., fossil fuels and other ores), (3) anthropogenic releases from production processes, and (4) remobilization of Hg from soils, sediments, and water from past anthropogenic releases. Whatever the original source of Hg entry into the environment, the final receptors of such emissions are the atmosphere, aquatic ecosystems, soils, and biota. The biogeochemical cycle of Hg is complex in that several environmental compartments and processes are involved in the cycle. Estimates of Hg emissions to the atmosphere show that natural sources of Hg (median value

2.5×10^9 g year^{-1}) have been surpassed by anthropogenic sources (median value 3.6×10^9 g year^{-1}) as of 1983; among the main natural sources of Hg are wind-borne soil particles, sea salt spray, volcanic activity, forest wildfires, and biogenic emissions (Nriagu 1989).

Estimates have also been made for the amount of Hg that is released from combusting or processing certain raw materials, such as coal and other fuels; in estimates made in 1995, Hg from global coal combustion accounted for 1474.5 t (77%) (Pirrone et al. 1996). Hg released from industrial production processes has been difficult to estimate, because releases of Hg from point sources or those associated with disposal of products and wastes are unavailable. US data for 1994/1995 show that 10–40% of all anthropogenic releases of Hg came from industrial production or processes (Lawrence 2000). Hg is also remobilized from soils, water, and sediments as a result of past anthropogenic releases. Some anthropogenic activities also result in Hg emissions; these include agriculture, deforestation, and dam construction, all of which can increase Hg releases to aquatic ecosystems and can eventually result in Hg being accumulated by biota (PNUMA 2005).

Acosta and Asociados (2001) reported that 31.293 t of Hg were emitted to the atmosphere in Mexico in 1999. The main sources of annual Hg emissions were mining and refining of gold (11.27 t; 36.0% of the total), mining and refining of Hg (9.666 t; 30.8%), chloralkali plant processes (4.902 t; 15.7%), copper smelting (1.543 t; 4.9%), residential combustion of wood (1.168 t; 3.7%), carboelectric plants (0.7855 t; 2.5%), and oil refining (0.680 t; 2.2%). Other Hg emission sources (e.g., thermoelectrical plants, lead and zinc smelting, fluorescent lamps, and dental amalgams) accounted for 0.9413 t (3.0%).

3   Mercury in the Atmospheric Compartment

There are large fluctuations of Hg levels in the atmosphere. In addition, there are Hg exchanges between the atmosphere and terrestrial surfaces that are subjected to Hg deposition and subsequent re-emission (Gustin et al. 2008). Moreover, Hg levels are high in some geological belts of the earth, as well as in areas that have high Fe and Zn mineralization (Rytuba 2003). Transportation of pollutants (including Hg) by wind can affect surrounding areas more than source sites (Chopin and Alloway 2007; López et al. 2008). Different forms of Hg exist in the atmosphere. These Hg forms include gaseous elemental Hg (GEM), divalent reactive gaseous Hg (RGM), and particulate Hg (PHg) (Fu et al. 2010). Unlike other trace metals that are mainly found adsorbed or absorbed to the particulate phase in the atmosphere, most Hg (>95%) is in the gaseous phase (Aspmo et al. 2006; Valente et al. 2007). In atmospheric studies, total Hg is usually reported in the form of total gaseous mercury (TGM), which is equivalent to GEM plus RGM. In Mexico, studies related to Hg presence in the atmosphere are scarce, although we summarize those data that are available in Table 1. In this table, we include TGM values that were reported in studies from other sites so as to provide a general view of Hg levels that exist in the atmospheric compartment.

Table 1 Levels (ng m⁻³) of total gaseous Hg (TGM) in the atmosphere in Mexico and selected other areas worldwide

Site	TGM	Remarks	Reference
Zacatecas (Mexico)	71.82	Mining wastes are used for brick manufacturing	De la Rosa et al. (2004)
Mexico City (Mexico)	9.81	Storing of steel and food processing in the surroundings	De la Rosa et al. (2004)
Hidalgo (Mexico)	1.32	Agricultural activities in the area	De la Rosa et al. (2004)
Oaxaca (Mexico)	1.46	Tropical forests and no industrial activity	De la Rosa et al. (2004)
Mexico City (Mexico)	7.26	Urbanized area	Rutter et al. (2009)
Mexico City (Mexico)	5.0	Rural area	Rutter et al. (2009)
East China Sea (China)	1.52	Cruise between Shanghai and Bering Sea	Kang and Xie (2011)
Mace Head (Ireland)	1.41	Study site located in a clean sector	Ebinghaus et al. (2011)
Guizhou (China)	9.3–40,000	Abandoned Hg mine, artisanal Hg mining	Li et al. (2011)

TGM total gaseous mercury

The TGM values given in Table 1 vary by orders of magnitude; the lowest levels (1.32 ng m⁻³) corresponded to TGM from Hidalgo state, a rural area primarily engaged in agriculture; such TGM concentrations were comparable to those reported in samples taken during a cruise between Shanghai and the Bering Sea (1.52 ng m⁻³), and in a pristine site (1.41 ng m⁻³) in Ireland. The highest TGM value (71.82 ng m⁻³) in Mexico existed in Zacatecas, where rustic brick manufacturers use mining wastes as a raw material. In another study performed in Yinqiangou (China), TGM levels ranged from 9.3 to 40,000 ng m⁻³. The latter concentration was found to exist at a smelting site. In China, the maximum occupational standard for atmospheric Hg is set at 10 μg m⁻³. The reported value for Yinqiangou (9.3–40,000 ng m⁻³) exceeded this limit; such elevated TGM levels have been reported to exist in artisanal mining workers in China (Li et al. 2008).

4 Mercury in the Aquatic Environment

4.1 Coastal Sediments

Mercury is a dynamic element, and its chemical behavior in waters, sediments, and soils is complex and is influenced by several factors. These factors include redox, pH, salinity, alkalinity, hardness, and organic matter (i.e., composition, reactivity, concentration, etc.). As with other metals, sediments and soils serve as the main reservoirs for Hg; consequently, the levels, distribution, and speciation of Hg in these media must be established to understand its complex behavior. In Table 2, we present selected data on Hg levels that exist in coastal, marine, lagoonal, and estuarine sediments from distinct regions of Mexico. These data show that sediments

Table 2 Mercury concentrations (range in µg g^{-1} on a dry wt basis) in sediments from Mexican coasts

Location	Hg concentrations	Reference
Northwest coast, B.C.	0.030–0.097	Gutiérrez-Galindo et al. (2007)
Todos Santos Bay, B.C.	0.011–0.063	Gutiérrez-Galindo et al. (2008)
Port of Ensenada, B.C.	0.58±0.23[a]	Carreón-Martínez et al. (2002)
Santispac Bight, B.C.	0.006–0.060	Leal-Acosta et al. (2010)
Mangrove lagoon, B.C.	0.015–0.233	Leal-Acosta et al. (2010)
La Paz lagoon, B.C.S.	0.01–0.05	Kot et al. (1999)
Beach sands, SR, B.C.S.	0.01–0.16	Kot et al. (2009)
Coastal sediments, SR, B.C.S.	0.01–0.35	Kot et al. (2009)
Oyster culture areas, Sonora	0.01–0.35	García-Rico et al. (2003)
Kun Kaak Bay, Sonora	0.05–0.15	García-Hernández et al. (2005)
Guaymas Bay, Sonora	0.34–2.25	Green-Ruiz et al. (2005)
Ohuira lagoon	0.03–0.30	Ruiz-Fernández et al. (2009)
Chiricahueto lagoon	<0.002–0.60	Ruiz-Fernández et al. (2009)
Urías lagoon	<0.002–0.28	Ruiz-Fernández et al. (2009)
	0.20–0.46	Jara Marini et al. (2008a)
Navidad lagoon, Colima	0.002–0.032	Willerer et al. (2003)
Tampamachoco lagoon, Ver.	0.011±0.005[a]	Rosas et al. (1983)
Mandinga lagoon, Ver.	0.028±0.012[a]	Rosas et al. (1983)
Alvarado lagoon, Ver.	0.028–0.091	Guentzel et al. (2007)
Coatzacoalcos estuary, Ver.	0.585–1.41	Ochoa et al. (1973)
	0.11–57.94	Báez et al. (1975)
	0.062–0.209	Botello and Páez-Osuna (1986)
	0.070–1.061	Ruelas-Inzunza et al. (2009)
Del Carmen lagoon, Tabasco	0.009±0.003[a]	Rosas et al. (1983)
Atasta lagoon, Campeche	<0.007	Rosas et al. (1983)

B.C. Baja California, *B.C.S.* Baja California Sur, *SR* Santa Rosalía, *Ver.* Veracruz
[a] Mean concentration

reflect the actual or potential impact of being near mining, industrialized areas, municipal wastewater outlets, and tectonic and geothermal areas.

In Table 2, we present data on the Hg levels that have been reported in sediments along Mexico's Pacific coastal areas or in the Gulf of Mexico. These levels ranged from 0.0002 to 57.94 µg g^{-1}. This range clearly reveals pristine or nearly pristine concentrations in locations along the northwest coasts of Baja California (B.C.), Todos Santos Bay, B.C., Santispac Bight, La Paz lagoon, the Navidad lagoon, and the Alvarado lagoon. In contrast, Guaymas Bay and Coatzacoalcos estuary registered levels that were relatively high. These higher levels reflect the influence of both anthropogenic and natural Hg pollutant sources. However, high levels were also found at locations where no data exist to show that Hg enrichment has occurred; such cases include sites in Sonora where oysters are cultured. Sadiq (1992) reviewed Hg in sediments from different regions of the world and established a value of 0.05 µg g^{-1} as the background Hg concentration that normally exists in uncontaminated sediments. In this context, surficial sediments from Coatzacoalcos estuary have both low (0.062 µg g^{-1}) and extremely high (57.94 µg g^{-1}) Hg concentrations (Table 2).

Another approach frequently used to evaluate the contamination of marine sediments is to contrast levels found at a selected location with general background levels; such background concentrations may be established by using the concentration of the metal in different media (e.g., soil; Fergusson 1990) that occurs in a particular region, or in sedimentary rocks (Turekian and Wedepohl 1961). In addition, Hg levels found may be contrasted with established threshold effect levels.

The NOAA (National Oceanic and Atmospheric Administration) SQR (Screening Quick Reference) provides tables that present preliminary screening concentrations for Hg and numerous other contaminants; the values in these tables are used to identify coastal-resource areas that may potentially be affected by contaminants (Buchman 2008). Long et al. (1995), in a classical study, proposed two guideline values, the effects range-low (ERL) and the effects range-median (ERM) for Hg, the values of which were 0.15 $\mu g\ g^{-1}$ and 0.71 $\mu g\ g^{-1}$, respectively. The maximum Hg levels in coastal sediments at various sites in Mexico exceeded ERL and ERM limits. The ERL was exceeded at the Port of Ensenada, in mangrove lagoons, beach sands and coastal sediments of Santa Rosalia, and in oyster culture areas of Sonora. The ERM was exceeded at Guaymas Bay and at Coatzacoalcos, among others. However, caution should be exercised in using such comparisons. O'Connor et al. (1998) concluded that guidelines that are based on bulk chemistry can provide useful triggers for further analysis, but should not be used alone as indicators of toxicity. In this context, Adams et al. (1992) indicated that the toxicity of different sediments may differ by a factor of 10 or more for the same metal concentration. In addition, the toxicity of metals in sediments is highly influenced by chemical speciation of the metal of interest, the sediment type, and organic matter content. Below, we describe the general characteristics of several studies in which Hg contamination was evaluated in coastal sediments at different locations in Mexico.

Only fairly recently have studies been performed on mercury residues in coastal sediments of Mexico. Kot et al. (1999) was the first to evaluate Hg levels along the Pacific coast; his studies were performed in La Paz lagoon, Baja California. This lagoonal system is located in the south-eastern portion of the Baja California peninsula (110.30 and 110.42°W; 24.10 and 24.19°N). The drainage area of the lagoon has a unique geological setting, in which a zone of tectonic crustal deformation is cut by active faults (Henry 1989). Eighty surface sediment samples were collected in the lagoon by Kot et al. (1999); stream sediments and soils associated with the margins of the lagoon were also collected. These authors indicated that, in general, the Hg levels found (Table 2) were relatively low, with variations depending on the region; the highest concentrations were found at sites associated with discharges from the wastewater treatment facility at La Paz city. The authors reported that the Hg concentrations found in the La Paz area were lower than typical background values for uncontaminated marine deposits; the results were also lower than the average abundance of this metal in sedimentary rocks (i.e., shales, 0.4 $\mu g\ g^{-1}$; Turekian and Wedepohl 1961) and soils (0.01–0.50 $\mu g\ g^{-1}$; Fergusson 1990).

In other studies, mercury concentrations were evaluated in surface sediments off the northwest coast of Baja California (116.8 and 117.3°W; 31.9 and 32.6°N) (Table 2), and results indicated a relatively homogenous spatial distribution of the metal. Hg enrichment occurred at four stations located in the northern and central zones. In general, Hg levels were relatively low and posed no environmental concerns, because the levels found were lower than the North American marine sediment quality guidelines of the NOAA's National Status and Trends Program for low and medium toxic effects of Hg in marine sediment (Long et al. 1995; Gutiérrez-Galindo et al. 2007).

Another study was performed in Northwest Mexico, where Hg levels in sediment samples from Todos Santos Bay (located 100 km south of the US–Mexico border) on the Pacific coast of the Baja California peninsula (116.6 and 116.85°W; 31.7 and 31.9°N) were analyzed (Table 2). This bay is exposed to potential pollution from domestic and industrial effluents, shipping and fishing traffic, and agricultural runoff (Gutiérrez-Galindo et al. 2008). With the exception of the port zone and marina, the bay was considered by these authors to be a region that has relatively low levels of trace metals. The average concentration of Hg found (0.023 µg g^{-1}) in Todos Santos Bay is similar to those in La Paz lagoon, Baja California Sur (0.020 µg g^{-1}).

Romero-Vargas (1995) studied sediments from the port of Ensenada (inside Todos Santos Bay) and discovered that they had relatively low metal concentrations, except for the zone encompassed by the port zone and marina (covering an area of 1.15 km^2), where metals tend to increase in fine-grain sediments from the effects of low wave energy. In addition to discharges from urban runoff and the contributions released from marine vessels, dry docks liberate antifouling paints and mine tailings from sand blasting operations (Carreón-Martínez et al. 2002). Each time that the port zone of Ensenada is dredged, from 150,000 to 300,000 m^3 of anoxic sediments are extracted and transported to other sites within and outside the Bay. Considering the concentration of Hg in such sediments (Table 2), one can estimate that approximately 45–96 kg of Hg is present in the sediment dredged during each operation. Carreón-Martínez et al. (2001, 2002) examined a core from the port zone to evaluate the importance of pyrite and other geochemical fractions as Hg reservoirs. They found that Hg in sediments was associated mainly with the Hg-HCl fraction (<0.03–0.17 µg g^{-1}) that includes the Hg linked to carbonates, Fe and Mn oxyhydroxides, iron monosulfides, and to the Hg-pyrite (<0.0.03–0.20 µg g^{-1}) linked to pyrite and other iron monosulfides.

Leal-Acosta et al. (2010) examined the composition of sediments of the intertidal geothermal hot spring zone and the adjacent area of Playa Santispac in Bahia Concepción (pristine area), Baja California (111.88 and 111.89°W; 26.75 and 26.77°N) (Table 2). High concentrations of Hg (and As) were found in the sediments of the geothermal sources. Hg levels decreased rapidly in the adjacent small mangrove lagoon sediments and finally reached background concentrations of 0.006–0.060 µg g^{-1}. The authors concluded that the geothermal hot springs located in Playa Santispac are important sources for imparting Hg (and As) to the sediments of the mangrove lagoon, and to the adjacent part of Concepción Bay. However, due to decreased temperature and oxygenated sea water, Hg is rapidly incorporated in the solid phase near the hot springs, and probably also into the freshly forming Mn-oxides and silicates.

Kot et al. (2009) traced the halo of Hg's dissipation in copper mines under arid conditions, at El Boleo mining district near Santa Rosalía in east-central Baja California, Mexico (27°24'–27°40'N and 112°22'–112°24'W). In this region, copper mining and smelting operations were abundant during the period 1885–1985. They found that marine sediments near Santa Rosalía are affected by material discharged through creeks and represent an indicator of metal dissipation from the mines. The alluvium in downstream waters included waste products from the smelter. Kot et al. 2009 found evidence that Hg was dispersed from the abandoned mines via stream sediments. The halo of Hg contamination was of a local character. Except for the two sites found near the San Luciano mine and in the Providencia stream, Hg concentrations far from the San Luciano stream approached the regional average. Nevertheless, this background level (0.14 µg g⁻¹) was an order of magnitude greater than that of La Paz lagoon and their associated stream sediments (0.016–0.020 µg g⁻¹; Kot et al. 1999). Coastal sediments adjacent to the affected area were found to be Hg enriched by an order of magnitude, compared to the smaller values found in the carbonate-diluted sediments from the south of Santa Rosalía. The highest Hg concentrations were registered in the harbor, probably from slag material originating from the smelter. The authors concluded that, in such an area, Hg is dispersed in the environment by diverse mechanisms: (a) the migration of Hg, in and with weathered material washed down from the immediate area of the abandoned mines; (b) oxidation of Hg sulfides typically resulting in the formation of soluble species (organic and chloride complexes). Such Hg compounds are easily trapped and immobilized by co-precipitation with iron oxides and amorphous aluminum oxides. Considering the arid climate, transport of particulate forms of Hg appears to be more significant, since the solubility of Hg sulfides is extremely low.

A peculiar study was performed in 2003 at Kun Kaak Bay, Sonora (112°0.4′W and 28°52′N) by García-Hernández et al. (2005). They reported levels of Hg and other trace metals in sediments and organisms during a harmful algal bloom. This event was associated with a massive die-off of fish and mollusks. An analysis of phytoplankton showed the presence of *Chatonella marina*, *Gymnodinium catenatum*, and *Gymnodinium sanguineum*. The levels of Hg in sediments (Table 2) were higher than background levels for this area.

Monthly measurements of Hg and other trace metals were made by García-Rico et al. (2003, 2006) in intertidal surface sediments from four areas, in which oysters were cultured. The four areas were near Puerto Peñasco (site 1, 113°23.1′W and 31°11.2′N), Caborca (site 2, 113°27.6′W and 31°10.3′N), Hermosillo (site 3, 1,110°55.5′W and 28°49.5′N), and Guaymas (site 4, 110°58.0′W and 27°54.5′N), which lie along the Sonoran coast (Table 2). The mean Hg concentration registered at the four sites was 0.07 µg g⁻¹, with the highest level found at site 3 (0.35 µg g⁻¹). The lowest mean levels were detected in February and the highest in August. The authors concluded that Hg levels were within typical background concentrations for uncontaminated marine deposits. Considering the examined geochemical fractions in surficial sediments, the authors found that Hg was detected only in the exchangeable and residual fractions, with the highest levels in the residual fraction. The first fraction was attributed to past mining activities and to residues from the atmospheric route. The residual fraction is probably linked to the structure of crystalline minerals and to unavailable unreactive forms of the sediments.

Mercury concentrations in surface sediments from Guaymas Bay (Sonora) (110°48′ and 110°55′W; 27°50′ and 28°00′W) were investigated by Green-Ruiz et al. (2005). The authors found an average Hg level of 1.0±0.5 µg g⁻¹, which is higher than the values reported for the coastal zone of Baja California (Table 2). In this study, the highest concentrations were found in Guaymas city, near the shipyards, and areas where the canning industry and the fishing fleet are located. According to the authors, Guaymas Bay has higher levels than those reported in other coastal ecosystems that are considered to be unpolluted or moderately polluted, such as La Paz lagoon (Kot et al. 1999), Todos Santos Bay (Gutiérrez-Galindo et al. 2008), and the Northwest coast in Baja California (Gutiérrez-Galindo et al. 2007).

Ruiz-Fernández et al. (2009) determined Hg and other metals in four selected sediment cores collected in three lagoons: Ohuira (108.8°W; 25.6°N), Chiricahueto (107.5°W; 24.6°), and Urías (106.3°W; 23.1°N). These ecosystems are located in the coastal plain of the southeastern Gulf of California; they have been included among the Priority Zones of the National Commission for the Knowledge and Use of Biodiversity of Mexico (CONABIO 2004) because of their importance as threatened natural areas. Ohuira is a shallow and brackish coastal lagoon that supports local fishing, shrimp farming, and intensive agriculture in the surrounding areas. Chiricahueto lagoon is a wetland marsh that has a limited water exchange with the sea and a freshwater swamp area produced by wastewaters from Culiacán valley. Critical pollution by agrochemicals, municipal and industrial loads, and urban wastes have been reported for this ecosystem (Páez-Osuna et al. 2007). Urías lagoon is a shallow water body that has a free and permanent exchange with the sea. The section that lies close to the sea functions as a navigation channel for the Mazatlán harbor shipping terminal. This area is surrounded by human settlements; in the area far from the mouth of the water body there is a thermoelectric plant and two shrimp farms (400 ha in pool size). The levels of Hg found in the sediment cores of the three lagoons were of the same order of magnitude, although the highest concentrations were found in Chiricahueto (Table 2). Ages and sedimentation rates were calculated by using ^{210}Pb activities in the three cores. From such values Hg fluxes were estimated; although fluxes varied with the time, there was a clear tendency for increased Hg levels in recent years. In Ohuira, the increase was more evident. For example, in the years 1905–1945, the Hg accumulation fluctuated between 1 and 15 ng cm^{-2}year^{-1}, while in years 1985–2005 the Hg fluxes were between 15 and 31 ng cm^{-2}year^{-1}. The authors concluded that Chiricahueto and Urías showed consistent signs of Hg pollution, with enrichment factors from 10 to 80, respectively. They explained that the historical increase of Hg obtained from the sediment records was related to the release of agricultural wastes in Ohuira and Chiricahueto, and to atmospheric release of the exhausts generated by the thermoelectric plant of Mazatlán, located in Urías lagoon.

Jara Marini et al. (2008a) analyzed surface sediments, water samples, and biota in Urías lagoon to evaluate the concentrations of Hg. Measurements in surface sediments (0–2.5 cm) revealed moderate Hg levels that were relatively homogenous. These concentrations were comparable to those reported by Ruiz-Fernández et al. (2009) in a sediment core (Table 2). According to McDonald et al. (1996), the total metal concentrations determined by these authors were higher than the Threshold Effects Level (TEL), and Hg concentrations were slightly higher than the Probable Effects Level (PEL). In addition, Jara Marini et al. (2008a) determined the bioavailable fraction of Hg (operationally defined: reactive + pyrite; Huerta-Díaz and Morse

1990). They found levels between 0.11 and 0.15 µg g^{-1}, and these levels fell between the TEL and PEL criteria values for sediment quality, indicating a level that is occasionally expected to produce adverse biological effects.

Willerer et al. (2003) studied Hg levels in surface sediments of the Marabasco River, the estuary, and the lagoon (Navidad lagoon). The watershed of the Marabasco River has a tectonically active province, which is rich in mineral deposits, Hg-ore deposits, an abandoned Hg-ore mine (Mina Martínez), and the Peña-Colorado iron-ore mine in Sierra "El Mamey." The river originates in the Sierra Manantlán and supplies freshwater for irrigating a large agricultural area, and finally flows into Navidad lagoon. For Navidad lagoon (104.65° and 104.68°W; 19.18° and 19.20°N), the Hg content in surface sediments varied from 0.002 to 0.032 µg g^{-1} (Table 2), with a mean concentration of 0.015 ± 0.009 µg g^{-1}. These relatively low levels were regarded to be natural background levels in sedimentary material. The source of variation is attributed to the grain size of sediments and their organic matter content; finer particles were richer in organic matter and Hg content. The authors concluded that existing Hg deposits and the ancient Hg mines situated in the drainage basin of the Marabasco River were the sources of the excess Hg in the sediments of the Navidad lagoon.

Rosas et al. (1983) conducted a pioneering study in Mexico on the relative heavy metal pollution that exists in the following four coastal lagoons in the Gulf of Mexico: Tampamachoco (97.6°W; 20.8°N), Mandinga (96.1°W; 19.0°N), Del Carmen (93.9°W; 18.3°N), and Atasta (92.1°W; 18.6°N). The four lagoons are characterized by fishing activities, oil extraction (PEMEX) activities, and having rural residential communities in the watershed. Mandinga is an urbanized area, which is close to the port and the city of Veracruz. The authors established the levels of Hg and other metals that existed in water, oysters, and sediment samples from these lagoons. The highest Hg values in surface sediments were reported from Mandinga lagoon, and the lowest concentrations were reported from Atasta lagoon (Table 2). They concluded that the variations of Hg content in the sediments of the four lagoons may result from their different edaphic composition.

The Alvarado lagoon system located (95°45' and 96°00'W; 18°40' and 18°53'N) within the Papaloapan River in southern Veracruz is a shallow system that is connected to the Gulf of Mexico through a narrow channel. This lagoon is a large mangrove-dominated coastal wetland that is formed by the confluence of four rivers, the Papaloapan, Blanco, Acula, and Limón, which descend from the central Mexican cordillera. The main economic activities of the watershed associated with this lagoon system are agriculture, including sugarcane cultivation and cattle ranching, and fishing. The nearest urban areas are the cities of Veracruz (50 km distant), Oaxaca (250 km distant), and Puebla (300 km distant). The potential sources of Hg in the area comprise wastes from fisheries and aquaculture, agriculture, and urban wastewater (population ~50,000 inhabitants). Guentzel et al. (2007) collected water, fish, sediment, and hair samples during the wet and dry seasons of 2005 and analyzed the samples for Hg content. The authors described the surface sediments of the Alvarado lagoon as being comprised of a mixture of sand, mud, and shells. The levels of Hg found were not significantly different between the wet and dry seasons of 2005 (Table 2). The authors concluded that the values found in the samples they collected were within the US EPA background sediment criteria level of <0.30 µg g^{-1} (US EPA 1997) and were below the threshold effects level for marine sediments (Buchman 1999).

Within the coastal zones of Mexico, Coatzacoalcos estuary (94°25′and 94°31′W; 17°46′and 18°10′N) is the location where more research on mercury, and perhaps also organic contaminants, has been carried out. The major petrochemical center of the Gulf of Mexico region is located in this estuary. From data included in Table 2, it is evident that Hg values found in the estuary were variable. Hg levels were more elevated during the decade of the 1970s, that is, Ochoa et al. (1973) and Báez et al. (1975) found mean values for Hg of 1.10 μg g⁻¹ and 8.31 ± 14.64 μg g⁻¹, respectively. During the decade of the 1980s, Botello and Páez-Osuna (1986) reported a mean value of 0.13 ± 0.07 μg g⁻¹. Finally, during 2005 and 2006 Ruelas-Inzunza et al. (2009) found Hg levels in the range of 0.070–1.061 μg g⁻¹. From such studies, it is clear that the high levels of Hg in Coatzacoalcos appear to be limited to critical zones of the estuary that are more heavily polluted. Residue level variability also depended on several factors related to industrial discharges: magnitude, time, concentration, chemical forms, tendency of Hg to accumulate, and characteristics of sediments (grain size, organic matter, and mineralogy).

Recently, Ruelas-Inzunza et al. (2009) studied Hg concentrations in surface sediments of the Coatzacoalcos River. Results showed that Hg levels ranged from 0.07 μg g⁻¹ at upstream sites that were far from industrialized areas to 1.06 μg g⁻¹ in the river area located near the highly industrialized port of Coatzacoalcos (Table 2). Báez et al. (1975) earlier found Hg levels of 0.11–57.94 μg g⁻¹ in surficial sediments from the same area. It can be seen that the highest Hg levels were detected at sites located near a refinery drainage (San Francisco stream) point, and along the most industrialized zone of Coatzacoalcos. When the Hg levels found at these sites are compared with levels found along other Mexican coastal areas, it is observed that the Coatzacoalcos estuary had higher Hg values than those at La Paz lagoon (Table 2), which was an unpolluted to moderately polluted site (Kot et al. 1999). However, the samples taken at the Coatzacoalcos River had Hg levels lower than those (14.2–31.4 μg g⁻¹) taken at Kastela Bay, Croatia, a site considered to be highly polluted (Kwokal et al. 2002). Considering the average Hg residue at Coatzacoalcos estuary, this ecosystem is (until 2008) regarded to be moderately contaminated at most sites tested.

4.2 Coastal Waters

The information on concentrations of Hg in waters from coastal environments in Mexico is limited to a few studies that are briefly summarized in this section. Some characteristics of Mexico's coastal water bodies have been described in the previous section. Historically, the locations at which more research has been conducted are Coatzacoalcos and the coastal lagoons of the Gulf of Mexico. In those coastal regions, such past research was motivated mainly by the existence of oil extraction and processing activities.

For the Urías lagoon, Jara Marini et al. (2008b) found that the concentrations of Hg in the dissolved fraction of waters ranged from 0.64 to 1.05 μg L⁻¹, while residue levels in the suspended fraction ranged from 2.22 to 2.64 μg L⁻¹. According to the authors, these concentrations exceeded the Hg levels commonly reported to exist in open sea and coastal waters as reported (Table 3) by Sadiq (1992), who gave a concentration of

Table 3 Mercury concentrations ($\mu g\ L^{-1}$) in waters sampled along Mexican coasts

Location	Hg concentrations	Reference
Urías lagoon, Sinaloa (dissolved)	0.64–1.05	Jara Marini et al. (2008b)
Urías lagoon, Sinaloa (suspended)	2.22–2.64	Jara Marini et al. (2008b)
Tampamachoco, Ver.	<0.2	Rosas et al. (1983)
Mandinga, Ver.	<0.2	Rosas et al. (1983)
Alvarado, Ver.	0.0009–0.0126[a]	Guentzel et al. (2007)
Coatzacoalcos estuary, Ver.	30.0±10.0[a]	Ochoa et al. (1973)
	0.1–75.0[a]	Báez et al. (1975)
	12.0±3.0[a]	Pérez-Zapata et al. (1984)
Mecoacán lagoon, Tabasco	0.1–0.8[a]	Pérez-Zapata et al. (1984)
Del Carmen lagoon, Tabasco	<0.2	Rosas et al. (1983)
	0.2–0.6[a]	Pérez-Zapata et al. (1984)
Machona lagoon, Tabasco	0.1–1.1[a]	Pérez-Zapata et al. (1984)
Atasta lagoon, Campeche	<0.2	Rosas et al. (1983)
Open sea and coastal waters	0.0002–1.42	Sadiq (1992)
Uncontaminated seawater	0.02	Sadiq (1992)

[a]Unfiltered water samples

0.02 $\mu g\ L^{-1}$ for uncontaminated seawater. In general, and in comparison to open seawaters, the Hg levels that exist in estuarine and coastal waters from various regions worldwide are high because these regions are proximate to input sources, both natural and anthropogenic. The values reported by Jara Marini et al. (2008b) in waters of the Urías lagoon are appreciably higher than the concentrations normally expected in uncontaminated seawaters. This may be related to the activities that take place in the area surrounding this lagoon (viz., shipping, fish industry, and food industry), domestic effluents, and particularly to the presence of a thermoelectric plant.

Rosas et al. (1983) reported that Hg concentrations in water were below the limit of detection (<0.2 $\mu g\ L^{-1}$) in the coastal lagoons of the Gulf of Mexico: Tampamachoco, Mandinga, Del Carmen, and Atasta. Guentzel et al. (2007) reported levels of Hg from 0.9 to 12.6 ng L^{-1} in unfiltered water samples taken during 2005 and 2006, in the Alvarado lagoon system. The higher concentrations they reported were associated with river waters that were enriched with organic matter, or with elevated total suspended solids in the brackish mixing zones of the lagoon. Total Hg in the lagoonal waters was significantly ($p < 0.001$) correlated with suspended matter in the water column. The authors concluded that these levels exceeded the US EPA ambient surface water quality criteria (0.77–1.4 ng L^{-1}) (US EPA 2006).

Results on the levels of Hg in water samples from Coatzacoalcos estuary were compiled by Villanueva and Botello (1998); measurements were made by Ochoa et al. (1973), Báez et al. (1975), and Pérez-Zapata et al. (1984), and they reported rather elevated levels of Hg (Table 3). Hg concentrations were higher than those reported in the Alvarado lagoon and in other coastal lagoons of the Gulf of Mexico, such as Tampamachoco, Mandinga, Del Carmen, and Atasta. Rosas et al. (1983) reported levels of <0.2 $\mu g\ L^{-1}$. Pérez-Zapata (1981) also reported lower levels in Machona lagoon (0.42±0.31 $\mu g\ L^{-1}$), Mecoacan lagoon (0.34±0.21 $\mu g\ L^{-1}$), and Del Carmen lagoon (0.36±0.09 $\mu g\ L^{-1}$). When these Hg levels were compared with those in estuarine and coastal waters from other regions (0.0002–1.42 $\mu g\ L^{-1}$; Sadiq 1992),

Coatzacoalcos estuary had extremely elevated Hg levels during the 1970s. A possible explanation for such elevated Hg values may be that water samples in early studies were not filtered. The waters of this system are characterized by having a high suspended particulate load that corresponds with the sediment load in the river, the tidal regime, and the discharge of effluents from a nearby petrochemical complex.

Coatzacoalcos estuary hosts the largest petrochemical facilities in Mexico. It is worth noting that previous analyses (between 1973 and 1984) showed rather high residues (from 0.1 to 30.0 $\mu g\ L^{-1}$) of Hg, when compared with the lower concentrations that were reported from the 1990s (from 0.0009 to 2.64 $\mu g\ L^{-1}$). This difference may have resulted from less sensitive analytical techniques and a greater chance of having had sample contamination in the early studies (Sadiq 1992).

5 Mercury in Aquatic Biota

Hg residues in aquatic biota are of great concern because of their potential implications to human health. In addition, such residues may indicate the general condition of ecosystems. Biomonitoring involves monitoring individual organisms, parts thereof, or even a community of organisms for the presence of pollutants. The purpose of such monitoring is to provide data on the quantitative aspects of environmental quality (Markert et al. 2003). Apart from indicating the concentration of trace metals or other contaminants present, or the effects on the environment and organisms therein, the gathered data may also provide evidence of environmental stresses (e.g., desiccation, acidification, and eutrophication).

Anthropogenic activities such as mining, and municipal and industrial discharges are usually the main sources of metals in the environment; metals can accumulate in aquatic ecosystems to toxic levels and may induce adverse effects that produce ecological concern (Wang et al. 2002). Diverse authors have used a wide range of organisms in biomonitoring studies. Oysters and mussels have been preferred over other species because of their biological characteristics. Despite the growing environmental concern for the presence of Hg in the environment, few monitoring studies on it have been performed in Mexico, a country that hosts a significant array of anthropogenic activities that may produce metal contaminants.

5.1 Invertebrates

Table 4 shows that *Vesicomya gigas* from Guaymas basin (hydrothermal field at 2,000 m depth) has a higher content of Hg than other bivalve species collected along the NW coastal zone of Mexico, or in other areas worldwide. These Hg residues in the Guaymas basin probably resulted from natural input sources of Hg that were concentrated in this organism from bioaccumulation. Comparing Hg content in *Crassostrea corteziensis* ($0.02 \pm 0.003\ \mu g\ g^{-1}$), collected from the coast of Sonora, and *Crassostrea gigas* ($0.16 \pm 0.06\ \mu g\ g^{-1}$) from the Gulf of California with levels in those similar bivalve species (from 0.04 to 0.11), we found that values were similar.

Table 4 Concentration of mercury ($\mu g\ g^{-1}$ on a dry wt basis) in molluscs from different areas

Species	Area	Hg	Reference
National			
Vesicomya gigas	Guaymas basin, Gulf of California	2.41 ± 2.21	Ruelas-Inzunza et al. (2003b)
Crassostrea corteziensis	Sonoran coast	0.02 ± 0.003	García-Rico et al. (2010)
Crassostrea gigas	SE Gulf of California	0.16 ± 0.06	Osuna-Martínez et al. (2010)
International			
Crassostrea gigas	Deep Bay, Hong Kong	0.084 ± 0.19	Phillips et al. (1982)
Crassostrea sp.	Gulf of Paria, Venezuela	0.04 ± 0.02	Roja de Astudillo et al. (2002)
Perna viridis	Gulf of Paria, Venezuela	0.06 ± 0.05	Roja de Astudillo et al. (2002)
Crassostrea rhizophorae	Northeast Brazil	0.08 ± 0.05	Vaisman et al. (2005)
Mytilus galloprovincialis	NW Mediterranean	0.23 ± 0.15	Zorita et al. (2007)
Dreissena polymorpha	Ebro River, NE Spain	0.11 ± 0.14	Carrasco et al. (2008)

Table 5 Concentrations of mercury ($\mu g\ g^{-1}$ on a dry wt basis) in shrimp sampled from different areas

Species	Area	Tissue	Hg	Reference
National				
Farfantepenaeus brevirrostris	AEP lagoon, Sinaloa	Hepatopancreas	0.35 ± 0.07	Ruelas-Inzunza et al. (2004)
		Muscle	0.21 ± 0.07	Ruelas-Inzunza et al. (2004)
Farfantepenaeus californiensis	AEP lagoon, Sinaloa	Hepatopancreas	0.62 ± 0.11	Ruelas-Inzunza et al. (2004)
		Muscle	0.13 ± 0.08	Ruelas-Inzunza et al. (2004)
Litopenaeus stylirostris	AEP lagoon, Sinaloa	Hepatopancreas	0.57 ± 0.01	Ruelas-Inzunza et al. (2004)
		Muscle	0.30 ± 0.036	Ruelas-Inzunza et al. (2004)
Litopenaeus vannamei	AEP lagoon, Sinaloa	Hepatopancreas	0.72 ± 0.07	Ruelas-Inzunza et al. (2004)
		Muscle	0.20 ± 0.01	Ruelas-Inzunza et al. (2004)
Xiphopenaeus kroyery	AEP lagoon, Sinaloa	Hepatopancreas	0.27 ± 0.04	Ruelas-Inzunza et al. (2004)
		Muscle	0.13 ± 0.04	Ruelas-Inzunza et al. (2004)
International				
Crangon crangon	Limfjord, Denmark	Muscle	0.09 ± 0.03	Riisgard and Famme (1986)
Penaeus sp.	Malaysia	Muscle	0.36 ± 0.13	Rahman et al. (1997)
Penaeus semisulcatus	Gulf of Arabia	Whole tissue	0.013 ± 0.007	Al-Saleh and Al-Doush (2002)
Penaeus semisulcatus	Northern Persian Gulf	Muscle	0.19 ± 0.05	Elahi et al. (2007)
Litopenaeus stylirostris	New Caledonia	Muscle	0.20 ± 0.06	Chouvelon et al. (2009)

The exception was the $0.23 \pm 0.15\ \mu g\ g^{-1}$ value reported by Zorita et al. (2007) in *Mytilus galloprovincialis* collected from the NW Mediterranean, which has been previously reported to be an area that has significant natural sources of mercury, as well as anthropogenic discharges.

The Hg content that exists in marine crustaceans collected from Mexico and from other areas is presented in Table 5. Only Ruelas-Inzunza et al. (2004) have performed studies on shrimp; they sampled the Altata-Ensenada del Pabellón (AEP)

coastal lagoon, which is associated with drainage of an agriculturalized basin. In their study, the hepatopancreas of shrimp showed higher Hg values than did muscle (0.27–0.72 µg g^{-1} vs. 0.13–0.30 µg g^{-1}, respectively). The higher residue values present in the hepatopancreas probably relate to the biological functions performed by this organ (viz., metabolize xenobiotics, digest food, store lipids and carbohydrates, and synthesize enzymes and proteins) (Manisseri and Menon 1995). Values reported in muscle of shrimp from AEP lagoon were higher than the mercury content found by Al-Saleh and Al-Doush (2002) in *Penaeus semisulcatus* from the Gulf of Arabia, but lower than the values reported by Rahman et al. (1997) in *Penaeus* sp. from Malaysia. Chouvelon et al. (2009) reported 0.20 µg g^{-1} in muscle of *Litopenaeus stylirostris* from New Caledonia. Ruelas-Inzunza et al. (2004) reported a level of 0.30 µg g^{-1} in the same species from the AEP lagoon (NW Mexico), indicating a similar contamination pattern as existed for mercury.

Clearly, more studies should be conducted along coastal Mexico, focusing on analyzing the Hg content in different marine organisms (viz., oysters, shrimp, and crabs), which are consumed at high rates, both locally and nationally.

5.2 Vertebrates

Vertebrates have been monitored in Mexico for Hg residue content. The monitored taxa included fish, reptiles, birds, and mammals. Data on the Hg concentrations in selected tissues of elasmobranchs is presented in Table 6. Reported levels of Hg were sourced from nine studies published between 1998 and 2012 and included 24 species. Data on elasmobranch species taken from the Pacific Ocean (seven contributions) were more abundant than those taken from the Gulf of Mexico area (two studies).

Tissues and organs of main interest were gills, brain, liver, pancreas, muscle, kidney, and fins. In most studies, Hg was analyzed in muscle tissue. Average Hg concentrations among species and tissues tested were highly variable. The highest Hg level (27.2 µg g^{-1}) reported was in the muscle of the smooth hammerhead shark *Sphyrna zygaena*, which was taken from the Gulf of California (García-Hernández et al. 2007). The lowest concentration was found in fins of the same species taken from offshore of Baja California Sur (Table 6). In a study with *S. zygaena* collected from the Ionian Sea, high total Hg (21.1 µg g^{-1} on a wet wt basis, equivalent to 70.3 µg g^{-1} on a dry wt basis) levels were found in muscle tissue (Storelli et al. 2003). The average Hg concentration detected in muscle of all species included in Table 6 was 2.59 ± 4.59 µg g^{-1}; in the species collected from the Gulf of Mexico the average was 1.98 ± 1.91 µg g^{-1}, and in species from the Pacific Ocean the average was 2.62 ± 4.72 µg g^{-1}. Results were highly variable and may have resulted from the heterogeneity of compared species, their sizes, and particular habit. For example, muscle tissue levels of Hg varied by two orders of magnitude (from 0.20 to 27.2 µg g^{-1}) in *S. zygaena* (three reports), and one order of magnitude (from 0.89 to 4.6 µg g^{-1}) in *Prionace glauca* (three reports) and *Carcharhinus falciformis* (from 0.99 to 3.4 µg g^{-1}; two reports).

The Hg levels in muscle tissue of teleost fish from the Pacific Ocean and the Gulf of Mexico are presented in Table 7. In Mexican waters, teleost fish have been more

Table 6 Mercury levels (µg g^{-1} dry wt) in selected tissues and organs of elasmobranchs from Mexican waters

Species	Common name	Tissue	Site	Hg	Reference
Rhizoprionodon terraenovae	Atlantic sharpnose shark	Gills	Gulf of Mexico (Veracruz state)	0.66	Núñez-Nogueira et al. (1998)
Rhizoprionodon terraenovae	Atlantic sharpnose shark	Brain	Gulf of Mexico (Veracruz state)	0.45	Núñez-Nogueira et al. (1998)
Rhizoprionodon terraenovae	Atlantic sharpnose shark	Liver	Gulf of Mexico (Veracruz state)	0.16	Núñez-Nogueira et al. (1998)
Rhizoprionodon terraenovae	Atlantic sharpnose shark	Pancreas	Gulf of Mexico (Veracruz state)	0.51	Núñez-Nogueira et al. (1998)
Rhizoprionodon terraenovae	Atlantic sharpnose shark	Muscle	Gulf of Mexico (Veracruz state)	0.63	Núñez-Nogueira et al. (1998)
Rhizoprionodon terraenovae	Atlantic sharpnose shark	Kidney	Gulf of Mexico (Veracruz state)	0.42	Núñez-Nogueira et al. (1998)
Carcharhinus leucas	Bull shark	Muscle	Altata-Ensenada del Pabellón (SE Gulf of California)	0.20	Ruelas-Inzunza and Páez-Osuna (2005)
		Liver	Altata-Ensenada del Pabellón (SE Gulf of California)	0.60	Ruelas-Inzunza and Páez-Osuna (2005)
Sphyrna lewini	Scalloped hammerhead	Muscle	Altata-Ensenada del Pabellón (SE Gulf of California)	4.84	Ruelas-Inzunza and Páez-Osuna (2005)
		Liver	Altata-Ensenada del Pabellón (SE Gulf of California)	0.12	Ruelas-Inzunza and Páez-Osuna (2005)
Carcharhinus limbatus	Blacktip shark	Muscle	Gulf of Mexico	3.33	Núñez-Nogueira (2005)
		Liver	Gulf of Mexico	7.78	Núñez-Nogueira (2005)
		Gills	Gulf of Mexico	7.03	Núñez-Nogueira (2005)
		Brain	Gulf of Mexico	1.33	Núñez-Nogueira (2005)
Sphyrna zygaena	Smooth hammerhead shark	Muscle	Off Baja California Sur, Mexico	0.73	Escobar-Sánchez et al. (2010)
		Fin	Off Baja California Sur, Mexico	0.007	Escobar-Sánchez et al. (2010)
Prionace glauca	Blue shark	Muscle	Off Baja California Sur, Mexico	4.6[a]	Escobar-Sánchez et al. (2011)
Sphyrna zygaena	Smooth hammerhead shark	Muscle	Gulf of California	27.2[a]	García-Hernández et al. (2007)
Alopias pelagicus	Pelagic thresher	Muscle	Gulf of California	4.3[a]	García-Hernández et al. (2007)
Rhizoprionodon longurio	Pacific sharpnose shark	Muscle	Gulf of California	4.3[a]	García-Hernández et al. (2007)
Carcharhinus obscurus	Dusky shark	Muscle	Gulf of California	3.82[a]	García-Hernández et al. (2007)
Sphyrna lewini	Scalloped hammerhead	Muscle	Gulf of California	3.56[a]	García-Hernández et al. (2007)
Nasolamia velox	Whitenose shark	Muscle	Gulf of California	3.36[a]	García-Hernández et al. (2007)
Carcharhinus limbatus	Blacktip shark	Muscle	Gulf of California	1.68[a]	García-Hernández et al. (2007)

Carcharhinus falciformis	Silky shark	Muscle	Gulf of California	0.99[a]	García-Hernández et al. (2007)
Prionace glauca	Blue shark	Muscle	Gulf of California	0.89[a]	García-Hernández et al. (2007)
Mustelus henlei	Brown smoothhound shark	Muscle	Gulf of California	0.59[a]	García-Hernández et al. (2007)
Triakis semifasciata	Leopard shark	Muscle	Gulf of California	0.26[a]	García-Hernández et al. (2007)
Prionace glauca	Blue shark	Muscle	Baja California Peninsula	1.96	Maz-Courrau et al. (2012)
Carcharhinus falciformis	Silky shark	Muscle	Baja California Peninsula	3.40	Maz-Courrau et al. (2012)
Isurus oxyrinchus	Mako shark	Muscle	Baja California Peninsula	1.05	Maz-Courrau et al. (2012)
Sphyrna zygaena	Smooth hammerhead shark	Muscle	Baja California Peninsula	0.98	Maz-Courrau et al. (2012)
Sphyrna lewini	Scalloped hammerhead	Muscle	Sinaloa state	2.72[a]	Hurtado-Banda et al. (2012)
Sphyrna lewini	Scalloped hammerhead	Liver	Sinaloa state	0.83[a]	Hurtado-Banda et al. (2012)
Rhizoprionodon longurio	Pacific sharpnose shark	Muscle	Sonora state	3.05[a]	Hurtado-Banda et al. (2012)
Rhizoprionodon longurio	Pacific sharpnose shark	Liver	Sonora state	0.21[a]	Hurtado-Banda et al. (2012)
Mustelus albipinnis	Smoothhound shark	Muscle	Sonora state	1.12[a]	Hurtado-Banda et al. (2012)
Mustelus albipinnis	Smoothhound shark	Liver	Sonora state	0.42[a]	Hurtado-Banda et al. (2012)
Dasyatis longus	Longtail stingray	Muscle	Gulf of California	2.34[a]	García-Hernández et al. (2007)
Dasyatis brevis	Whiptail stingray	Muscle	Gulf of California	1.48[a]	García-Hernández et al. (2007)
Rhinoptera steindachneri	Pacific cownose ray	Muscle	Gulf of California	1.42[a]	García-Hernández et al. (2007)
Rhinoptera steindachneri	Pacific cownose ray	Muscle	Upper Gulf of California	0.31	Gutiérrez-Mejía et al. (2009)
Rhinoptera steindachneri	Pacific cownose ray	Liver	Upper Gulf of California	0.22	Gutiérrez-Mejía et al. (2009)
Rhinobatos productus	Shovelnose guitarfish	Muscle	Gulf of California	1.02[a]	García-Hernández et al. (2007)
Gymnura marmorata	California butterfly ray	Muscle	Gulf of California	0.46[a]	García-Hernández et al. (2007)
Rhinobatos glaucostigma	Speckled guitarfish	Muscle	Gulf of California	0.46[a]	García-Hernández et al. (2007)
Narcine entemedor	Giant electric ray	Muscle	Gulf of California	0.39[a]	García-Hernández et al. (2007)
Zapteryx exasperata	Banded guitarfish	Muscle	Gulf of California	0.36[a]	García-Hernández et al. (2007)
Myliobatis californica	Bat Eagle ray	Muscle	Gulf of California	0.17[a]	García-Hernández et al. (2007)

[a]Original results were given on a wet wt basis; conversions to dry wt were made based on a 70% water content

Table 7 Mercury levels (μg g^{-1} dry wt) in muscle tissue of teleost fish collected from Mexican waters

Species	Common name	Site	Hg	Reference
Istiophorus platypterus	Indo-Pacific sailfish	Gulf of California	1.32[a]	García-Hernández et al. (2007)
Makaira mazara	Indo-Pacific blue marlin	Gulf of California	1.18[a]	García-Hernández et al. (2007)
Acanthocybium solandri	Wahoo	Gulf of California	0.49[a]	García-Hernández et al. (2007)
Tetrapturus audax	Striped marlin	Gulf of California	0.46[a]	García-Hernández et al. (2007)
Seriola lalandi	Yellowtail amberjack	Gulf of California	0.23[a]	García-Hernández et al. (2007)
Coryphaena hippurus	Pompano dolphinfish	Gulf of California	0.16[a]	García-Hernández et al. (2007)
Thunnus albacares	Yellowfin tuna	Gulf of California	0.09[a]	García-Hernández et al. (2007)
Mycteroperca jordani	Gulf grouper	Gulf of California	1.18[a]	García-Hernández et al. (2007)
Mycteroperca rosacea	Leopard grouper	Gulf of California	1.12[a]	García-Hernández et al. (2007)
Epinephelus analogus	Spotted grouper	Gulf of California	0.89[a]	García-Hernández et al. (2007)
Paralabrax auroguttatus	Goldspotted sand bass	Gulf of California	0.56[a]	García-Hernández et al. (2007)
Mugil curema	White mullet	Veracruz, Veracruz	0.04	Reimer and Reimer (1975)
Sphyraena guachancho	Guachanche barracuda	Villa Cardel, Veracruz	0.11	Reimer and Reimer (1975)
Polynemus virginicus	Barbu	Villa Cardel, Veracruz	0.07	Reimer and Reimer (1975)
Centropomus sp.	Snook	Coatzacoalcos, Veracruz	0.27	Reimer and Reimer (1975)
Mugil cephalus	Gray mullet	Tampico, Tamaulipas	0.07	Reimer and Reimer (1975)
Mugil cephalus	Gray mullet	Guaymas, Sonora	0.03	Reimer and Reimer (1975)
Mugil cephalus	Gray mullet	Topolobampo, Sinlaoa	0.03	Reimer and Reimer (1975)
Mugil cephalus	Gray mullet	Mazatlán, Sinaloa	0.06	Reimer and Reimer (1975)
Anisotremus interruptus	Burrito grunt	Mazatlán, Sinaloa	0.12	Reimer and Reimer (1975)
Diapterus sp.	White mojarra	Ciudad del Carmen, Campeche	0.06	Reimer and Reimer (1975)
Thunnus albacares	Yellowfin tuna	Baja California Sur	0.51	Ordiano-Flores et al. (2011)
Not determined	Tuna (canned in oil)	–	4.06[a]	Velasco-González et al. (2001)
Not determined	Sardine (canned in oil)[b]	–	2.44[a]	Velasco-González et al. (2001)
Thunnus albacares	Yellowfin tuna (canned in oil)	Sold in NW Mexico	0.258	Ruelas-Inzunza et al. (2011b)
Thunnus albacares	Yellowfin tuna (canned in water)	Sold in NW Mexico	0.362	Ruelas-Inzunza et al. (2011b)
Istiophorus platypterus	Indo-Pacific sailfish	Southeast Gulf of California	4.88[a]	Soto-Jiménez et al. (2010)
Tetrapturus audax	Striped marlin	Southeast Gulf of California	5.67[a]	Soto-Jiménez et al. (2010)
Tilapia mossambica	Tilapia	Baja California	0.05	Gutiérrez-Galindo et al. (1988)
Ariopsis felis	Catfish	Southern Gulf of Mexico	0.083	Vázquez et al. (2008)

Species	Common name	Location	Value	Reference
Oreochromis sp.	Tilapia	Coatzacoalcos estuary	0.054	Ruelas-Inzunza et al. (2009)
Gobiomorus polylepis	Finescale sleeper	Coatzacoalcos estuary	0.492	Ruelas-Inzunza et al. (2009)
Gerres cinereus	Yellowfin mojarra	Coatzacoalcos estuary	0.135	Ruelas-Inzunza et al. (2009)
Centropomus viridis	White snook	Coatzacoalcos estuary	0.612	Ruelas-Inzunza et al. (2009)
Lepisosteus osseus	Longnose gar	Coatzacoalcos estuary	0.141	Ruelas-Inzunza et al. (2009)
Elops affinis	Pacific ladyfish	Sinaloa coast	0.97	Ruelas-Inzunza et al. (2008)
Sphoeroides annulatus	Bullseye puffer	Sinaloa coast	0.77	Ruelas-Inzunza et al. (2008)
Lutjanus colorado	Colorado snapper	Sinaloa coast	0.53	Ruelas-Inzunza et al. (2008)
Diapterus peruvianus	Peruvian mojarra	Sinaloa coast	0.58	Ruelas-Inzunza et al. (2008)
Haemulopsis axillaris	Yellowstripe grunt	Sinaloa coast	1.18	Ruelas-Inzunza et al. (2008)
Pomadasys leuciscus	White grunt	Topolobampo	0.95	Ruelas-Inzunza et al. (2008)
Caranx caninus	Pacific crevalle jack	Topolobampo	3.32	Ruelas-Inzunza et al. (2008)
Oligoplites saurus	Leatherjacket	Topolobampo	1.74	Ruelas-Inzunza et al. (2008)
Centropomus armatus	Armed snook	Topolobampo	1.51	Ruelas-Inzunza et al. (2008)
Scomberomorus sierra	Pacific sierra	Topolobampo	0.64	Ruelas-Inzunza et al. (2008)
Roncador stearnsii	Spotfin croaker	Topolobampo	1.39	Ruelas-Inzunza et al. (2008)
Haemulopsis axillaris	Yellowstripe grunt	Topolobampo	1.01	Ruelas-Inzunza et al. (2008)
Nematistius pectoralis	Roosterfish	Topolobampo	1.34	Ruelas-Inzunza et al. (2008)
Paralichthys woolmani	Speckled flounder	Topolobampo	0.68	Ruelas-Inzunza et al. (2008)
Centropomus nigrescens	Black snook	Topolobampo	0.56	Ruelas-Inzunza et al. (2008)
Haemulon sexfasciatum	Graybar grunt	Santa María	1.49	Ruelas-Inzunza et al. (2008)
Trachinotus paitensis	Paloma pompano	Santa María	1.42	Ruelas-Inzunza et al. (2008)
Centropomus medius	Blackfin snook	Santa María	0.82	Ruelas-Inzunza et al. (2008)
Caulolatilus princeps	Ocean whitefish	Altata	0.57	Ruelas-Inzunza et al. (2008)
Pomadasys branickii	Sand grunt	Fishing ground	1.17	Ruelas-Inzunza et al. (2008)
Mugil curema	White mullet	Sinaloa coast	0.47	Ruelas-Inzunza et al. (2008)
Mugil cephalus	Striped mullet	Sinaloa coast	0.07	Ruelas-Inzunza et al. (2008)
Gerres cinereus	Yellow fin mojarra	Sinaloa coast	0.82	Ruelas-Inzunza et al. (2008)
Selar crumenophthalmus	Bigeye scad	Topolobampo	0.65	Ruelas-Inzunza et al. (2008)
Arius platypogon	Cominate sea catfish	Topolobampo	0.98	Ruelas-Inzunza et al. (2008)
Eucinostomus gracilis	Graceful mojarra	Fishing ground	0.68	Ruelas-Inzunza et al. (2008)

[a]Original results were given on a wet wt basis; conversions to dry wt were made based on a 70% water content
[b]Whole fish

studied than any other group. Eleven studies have been published (seven dealt with the Pacific Ocean and four dealt with other sites in Mexico), and these studies represent Hg levels for 47 species (species names were not provided in two studies). The first data on teleost fish Hg residues were published in 1975, and the newest data in 2011. As with elasmobranchs, the Hg levels in these fish varied by two orders of magnitude; the highest value (5.67 µg g^{-1} dry wt) was in the striped marlin (*Tetrapturus audax*) collected from the SE Gulf of California. The lowest Hg concentration (0.03 µg g^{-1} dry wt) was detected in the gray mullet (*Mugil cephalus*) taken from Guaymas, Sonora, and Topolobampo, Sinaloa.

Studies related to the occurrence of Hg in tissues of teleost fish are abundant, and geographic areas where levels of this metal in muscle exceed regulations for human consumers include Italy, Spain, Taiwan, Florida, and Oregon (Eisler 2010). It is notable that elevated concentrations of Hg do exist in top predator fish like marlins. Elevated Hg levels in muscle tissue were reported in the blue marlin (*Makaira nigricans*; 10.5 µg g^{-1} wet wt) collected from the northern Gulf of Mexico (Cai et al. 2007) and in the black marlin (*Makaira indica*) from NE Australia (Mackay et al. 1975).

Hg residue data on reptiles, birds, and mammals are presented in Table 8. Only four studies on reptiles in Mexico have been published, and all of these were on marine turtles. The levels found in these turtles ranged from 0.795 µg g^{-1} dry wt in liver of *Lepidochelys olivacea* to 0.0006 µg g^{-1} dry wt in blood of the same species. When Hg was measured in liver and kidney, these organs displayed the highest concentrations. It has been observed that in marine reptiles Hg seldom exceeds 0.5 µg g^{-1} dry wt (Eisler 2010); such is the case also with reptiles sampled from Mexico's ecosystems (with the exception of liver samples of *L. olivacea*).

Hg levels in birds have been studied more often than levels in reptiles and mammals. The only published study that deals with the Hg distribution in tissues of birds from Mexico compared levels between migratory and resident avifauna. For resident birds, the sequence of Hg concentrations detected in different tissues was as follows: liver > feathers > heart > muscle > viscera. For migratory birds, the descending order of Hg levels was feathers > liver > muscle > heart > viscera. The highest values for Hg in the liver (5.08 µg g^{-1} dry wt) were recorded in the olivaceous cormorant (*Phalacrocorax olivaceus*); the highest Hg concentration in feathers (3.57 µg g^{-1} dry wt) was detected in the American avocet (*Recurvirostra americana*). It has been established that Hg levels of 5.0 µg g^{-1} dry wt in feathers may cause adverse reproductive effects (Eisler 1987). Liver levels above 3.0 µg g^{-1} dry wt may cause sublethal effects to these birds (Hui et al. 2001).

In Mexico, marine mammals have been scarcely studied for their Hg content, although species that were stranded in the southern portion of the Gulf of California were examined for Hg residues (Ruelas-Inzunza et al. 2003a). The Hg concentrations detected in these species were highly variable, ranging from 70.35 µg g^{-1} dry wt in liver of the spinner dolphin (*Stenella longirostris*) to 0.145 µg g^{-1} dry wt in muscle of the gray whale (*Eschrichtius robustus*). It appeared that Hg levels did not play a key role in the stranding of these organisms. In a study with *S. longirostris* from the Lesser Antilles (Gaskin et al. 1974), Hg concentrations in liver were of the same magnitude as reported in the study of Ruelas-Inzunza et al. (2003a). To our

Table 8 Mercury levels ($\mu g \ g^{-1}$ dry wt) in selected tissues of reptiles, birds, and marine mammals from Mexico

Species	Common name	Tissue	Site	Hg	Reference
Reptiles					
Lepidochelys olivacea	Olive ridley turtle	Yolk	Oaxaca	0.028	Páez-Osuna et al. (2011)
		Albumen	Oaxaca	0.0010	Páez-Osuna et al. (2011)
		Eggshell	Oaxaca	0.0087	Páez-Osuna et al. (2011)
		Blood	Oaxaca	0.0006	Páez-Osuna et al. (2011)
Caretta caretta	Loggerhead turtle	Blood	Baja California Sur	<0.01	Ley-Quiñónez et al. (2011)
Chelonya midas	Green turtle	Fat	Baja California Sur	0.011	Kampalath et al. (2006)
		Liver	Baja California Sur	0.168	Kampalath et al. (2006)
		Muscle	Baja California Sur	0.059	Kampalath et al. (2006)
		Kidney	Baja California Sur	0.310	Kampalath et al. (2006)
Chelonia midas agassizii	Black sea turtle	Scutes	Gulf of California	0.17[a]	Presti et al. (1999)
Caretta caretta	Loggerhead turtle	Fat	Baja California Sur	0.028	Kampalath et al. (2006)
		Liver	Baja California Sur	0.183	Kampalath et al. (2006)
		Muscle	Baja California Sur	0.041	Kampalath et al. (2006)
		Kidney	Baja California Sur	0.135	Kampalath et al. (2006)
Lepidochelys olivacea	Olive ridley turtle	Fat	Baja California Sur	0.156	Kampalath et al. (2006)
		Liver	Baja California Sur	0.795	Kampalath et al. (2006)
		Muscle	Baja California Sur	0.144	Kampalath et al. (2006)
		Kidney	Baja California Sur	0.372	Kampalath et al. (2006)
Birds					
Pelecanus occidentalis	Brown pelican	Muscle	Altata-Ensenada del Pabellón Lagoon	2.11	Ruelas-Inzunza et al. (2007)
		Heart	Altata-Ensenada del Pabellón Lagoon	1.92	Ruelas-Inzunza et al. (2007)
		Liver	Altata-Ensenada del Pabellón Lagoon	3.70	Ruelas-Inzunza et al. (2007)
		Feathers	Altata-Ensenada del Pabellón Lagoon	2.05	Ruelas-Inzunza et al. (2007)
		Viscera	Altata-Ensenada del Pabellón Lagoon	0.85	Ruelas-Inzunza et al. (2007)

(continued)

Table 8 (continued)

Species	Common name	Tissue	Site	Hg	Reference
Phalacrocorax olivaceus	Olivaceous cormorant	Muscle	Altata-Ensenada del Pabellón Lagoon	1.13	Ruelas-Inzunza et al. (2007)
		Heart	Altata-Ensenada del Pabellón Lagoon	1.62	Ruelas-Inzunza et al. (2007)
		Liver	Altata-Ensenada del Pabellón Lagoon	5.08	Ruelas-Inzunza et al. (2007)
		Feathers	Altata-Ensenada del Pabellón Lagoon	3.19	Ruelas-Inzunza et al. (2007)
		Viscera	Altata-Ensenada del Pabellón Lagoon	1.02	Ruelas-Inzunza et al. (2007)
Casmerodius albus	Great egret	Muscle	Altata-Ensenada del Pabellón Lagoon	1.58	Ruelas-Inzunza et al. (2007)
		Heart	Altata-Ensenada del Pabellón Lagoon	1.67	Ruelas-Inzunza et al. (2007)
		Liver	Altata-Ensenada del Pabellón Lagoon	3.18	Ruelas-Inzunza et al. (2007)
		Feathers	Altata-Ensenada del Pabellón Lagoon	0.73	Ruelas-Inzunza et al. (2007)
		Viscera	Altata-Ensenada del Pabellón Lagoon	0.66	Ruelas-Inzunza et al. (2007)
Bubulcus ibis	Cattle egret	Muscle	Altata-Ensenada del Pabellón Lagoon	0.36	Ruelas-Inzunza et al. (2007)
		Heart	Altata-Ensenada del Pabellón Lagoon	0.32	Ruelas-Inzunza et al. (2007)
		Liver	Altata-Ensenada del Pabellón Lagoon	0.87	Ruelas-Inzunza et al. (2007)
		Feathers	Altata-Ensenada del Pabellón Lagoon	0.07	Ruelas-Inzunza et al. (2007)
		Viscera	Altata-Ensenada del Pabellón Lagoon	0.58	Ruelas-Inzunza et al. (2007)
Plegadis chihi	White-faced ibis	Muscle	Altata-Ensenada del Pabellón Lagoon	0.30	Ruelas-Inzunza et al. (2007)
		Heart	Altata-Ensenada del Pabellón Lagoon	0.42	Ruelas-Inzunza et al. (2007)
		Liver	Altata-Ensenada del Pabellón Lagoon	0.42	Ruelas-Inzunza et al. (2007)
		Feathers	Altata-Ensenada del Pabellón Lagoon	2.23	Ruelas-Inzunza et al. (2007)
		Viscera	Altata-Ensenada del Pabellón Lagoon	0.09	Ruelas-Inzunza et al. (2007)
Recurvirostra americana	American avocet	Muscle	Altata-Ensenada del Pabellón Lagoon	0.66	Ruelas-Inzunza et al. (2007)
		Heart	Altata-Ensenada del Pabellón Lagoon	0.40	Ruelas-Inzunza et al. (2007)
		Liver	Altata-Ensenada del Pabellón Lagoon	0.87	Ruelas-Inzunza et al. (2007)
		Feathers	Altata-Ensenada del Pabellón Lagoon	3.57	Ruelas-Inzunza et al. (2007)
		Viscera	Altata-Ensenada del Pabellón Lagoon	0.25	Ruelas-Inzunza et al. (2007)

Dendrocygna autumnalis	Black-bellied duck	Muscle	Altata-Ensenada del Pabellón Lagoon	0.023	Ruelas-Inzunza et al. (2007)
		Heart	Altata-Ensenada del Pabellón Lagoon	0.005	Ruelas-Inzunza et al. (2007)
		Liver	Altata-Ensenada del Pabellón Lagoon	0.20	Ruelas-Inzunza et al. (2007)
		Feathers	Altata-Ensenada del Pabellón Lagoon	0.38	Ruelas-Inzunza et al. (2007)
		Viscera	Altata-Ensenada del Pabellón Lagoon	0.10	Ruelas-Inzunza et al. (2007)
Anas cyanoptera	Cinnamon teal	Muscle	Altata-Ensenada del Pabellón Lagoon	0.264	Ruelas-Inzunza et al. (2007)
		Heart	Altata-Ensenada del Pabellón Lagoon	0.15	Ruelas-Inzunza et al. (2007)
		Liver	Altata-Ensenada del Pabellón Lagoon	0.15	Ruelas-Inzunza et al. (2007)
		Feathers	Altata-Ensenada del Pabellón Lagoon	0.49	Ruelas-Inzunza et al. (2007)
		Viscera	Altata-Ensenada del Pabellón Lagoon	0.07	Ruelas-Inzunza et al. (2007)
Fulica americana	American coot	Muscle	Altata-Ensenada del Pabellón Lagoon	0.163	Ruelas-Inzunza et al. (2007)
		Heart	Altata-Ensenada del Pabellón Lagoon	0.309	Ruelas-Inzunza et al. (2007)
		Liver	Altata-Ensenada del Pabellón Lagoon	0.77	Ruelas-Inzunza et al. (2007)
		Feathers	Altata-Ensenada del Pabellón Lagoon	1.18	Ruelas-Inzunza et al. (2007)
		Viscera	Altata-Ensenada del Pabellón Lagoon	0.23	Ruelas-Inzunza et al. (2007)
Marine mammals					
Eschrichtius robustus	Gray whale	Muscle	Gulf of California, Mexico	0.145	Ruelas-Inzunza et al. (2003a)
		Kidney	Gulf of California, Mexico	0.277	Ruelas-Inzunza et al. (2003a)
		Liver	Gulf of California, Mexico	0.185	Ruelas-Inzunza et al. (2003a)
Stenella longirostris	Spinner Dolphin	Muscle	Gulf of California, Mexico	1.66	Ruelas-Inzunza et al. (2000)
		Kidney	Gulf of California, Mexico	6.975	Ruelas-Inzunza et al. (2000)
		Liver	Gulf of California, Mexico	70.35	Ruelas-Inzunza et al. (2000)

[a]Original results were given on a wet wt basis; conversions to dry wt were made based on a 70% water content

Table 9 Mercury levels ($\mu g\ g^{-1}$ dry wt) in macrophytes and vestimentiferan tube worms from Mexico

Species	Tissue	Site	Hg	Reference
Macrophytes				
Codium amplivesciculatum	Fronds	Guaymas Bay, NW Mexico	0.099	Green-Ruiz et al. (2005)
Enteromorpha clathrata	Fronds	Guaymas Bay, NW Mexico	0.134	Green-Ruiz et al. (2005)
Gracilaria subsecundata	Fronds	Guaymas Bay, NW Mexico	0.095	Green-Ruiz et al. (2005)
Ulva lactuca	Fronds	Guaymas Bay, NW Mexico	0.058	Green-Ruiz et al. (2005)
Vestimentiferans				
Riftia pachyptila	Trophosome	Guaymas basin, NW Mexico	22.2	Ruelas-Inzunza et al. (2005)
	Vestimentum	Guaymas basin, NW Mexico	22.5	Ruelas-Inzunza et al. (2005)

knowledge, no other data in Mexico has been published on Hg residues in the whale (*E. robustus*); however, Varanasi et al. (1994) reported results on this whale species that had become stranded along the coasts of Washington, Alaska, and California. The average residues of Hg found in liver (0.27 $\mu g\ g^{-1}$ dry wt) and kidney (0.13 $\mu g\ g^{-1}$ dry weight) samples from the analyzed specimens (Varanasi et al. 1994) were comparable to those measured in specimens taken from Mexican waters.

5.3   Other Groups

Because of the dearth of Hg for biota, other than vertebrates and invertebrates, the few data available on other species are grouped for presentation in Table 9. Levels of Hg in macrophytes have been recorded only in specimens collected from NW Mexico; the concentrations detected in fronds of *Enteromorpha clathrata* were rather high (0.134 $\mu g\ g^{-1}$ dry wt). The Hg levels in *E. clathrata* were much higher than those found in the rest of the macroalgae species (Green-Ruiz et al. 2005). In a study with fronds of *Enteromorpha* spp., performed in NW Portugal, Leal et al. (1997) reported a high value of 0.160 $\mu g\ g^{-1}$ dry wt, which was comparable to values in the Green-Ruiz et al. (2005) study.

To our knowledge, only one study on Hg levels in vestimentiferan tube worms (*Rifita pachyptila*) has been published in Mexico. This study involved specimens collected from the hydrothermal field of the Guaymas basin at a 2,000 m depth. The Hg concentration in trophosome (the organ that hosts bacteria responsible for part of their metabolism) was 22.2 $\mu g\ g^{-1}$ dry wt. In *R. pachyptila* from other hydrothermal areas in the east Pacific Rise (13°N), Hg levels (from 11.3 to 23 $\mu g\ g^{-1}$ dry wt) in trophosome were comparable to values reported in specimens from the Guaymas basin. The vestimentum is a distinctive part of the body of *R. pachyptila*, made of muscle that forms a chitinous tube, where the animal lives. The average Hg concentration in the vestimentum of organisms from Guaymas basin was higher than those in *R. pachyptila* (from 1.5 to 4.0 μg g^{-1} dry wt) from the East Pacific Rise (13°N) (Cosson 1996). The elemental concentrations in organisms from vent sites were quite variable and resulted from variations in chemistry, intensity, frequency, and duration of vent fluids.

6 Effects in Humans

In the 1950s, mercury attracted attention as a potent environmental threat to human health because of the Minamata disease incident in Japan. In this event, residents of a small fishing village, where a plastic production plant was located, suffered many effects from organic mercury intake, including central nervous system (CNS) impairment in children, adults, and in utero exposed newborns. Studies in humans have demonstrated that the brain retains mercury for approximately 21 days, and this metal causes neurotoxicity when accumulated in the CNS (Echeverria et al. 2005). According to Landis and Yu (1999), the main effect of Hg is the inhibition of a large variety of enzymes, due to the affinity of this metal for SH groups. Methylmercury (CH_3Hg) is considered to be more toxic than inorganic mercury, since the methyl group increases its absorption into the bloodstream, enhances its bioavailability, and facilitates its distribution throughout the body (Friberg and Monet 1989).

Although fish contain omega-3 fatty acids and provide many benefits to humans who consume them, fish and other marine organisms are also potential dietary sources of metals and other pollutants such as mercury and methylmercury (Cohen et al. 2005). Xue et al. (2007) reported a positive relationship between consuming fish and the resulting mercury levels in human hair. Wang et al. (2002) pointed out that consuming fish and vegetables in Tianjin, China, posed a potential health risk to humans because both contained mercury.

In a Hg exposure study of a subsistence fishing community in western Mexico (Lake Chapala), it was found that consuming carp and other fish purchased or captured in Lake Chapala was associated with elevated Hg levels in hair (Trasande et al. 2010). The authors highlighted the fact that consuming these contaminated fish contributed significantly to hair Hg concentration in women of childbearing age, a sector of the population of high concern because of the potential for methylmercury-induced developmental neurotoxicity.

Ruelas-Inzunza et al. (2011a) evaluated the hazard quotient (HQ) for the NW Mexico population, based on the Hg content that existed in fish and shrimp they consumed. Though fish and shrimp consumption (9.01 kg person^{-1} year^{-1} and 1.49 kg person^{-1} year^{-1}, respectively) are lower than the world average, these authors found that organisms like *Caranx caninus* and the top predator *Sphyrna lewini* represented a potential health risk for consumers, mainly to fishermen and their families who consume two or three times more seafood than an average person. Levy et al. (2004) states that children are more vulnerable to the effects of mercury, and prolonged exposure may cause impairment of the developing of CNS, and motor function and behavioral disorders. In this context, White et al. (1992) pointed out that the symptoms of mercurialism are loss of mental capacities (i.e., memory, logical reasoning, or intelligence) and motor effects (i.e., imbalance, coordination, and tremor in muscles).

In humans, the kidney is a common target for metals, since it serves as a major excretory pathway for metals and has important metabolic functions. Nephrotoxic effects of Hg species are characterized by damage to cellular membrane and loss of mitochondrial functionality (Fowler 1996). People occupationally exposed to Hg have high urinary Hg levels (>50 μg L^{-1}) with the prospect of concomitant deleterious effects. Gonzalez-Ramírez et al. (1995) carried out a study in dental and non-dental personnel in Monterrey, Mexico, and found that urinary mercury levels were higher in personnel who formulated amalgams. Later, Woods et al. (2007) found a strong and positive correlation between urinary mercury levels and the number of amalgams emplaced and the elapsed time since their emplacement.

People living in proximity to mining areas are often as vulnerable to Hg exposures as are those occupationally exposed to Hg. Therefore, Acosta-Saavedra et al. (2011) concluded that women and children in Mexico face Hg exposure risks for both geographical and socioeconomical conditions. For example, in a monitoring study of Hg and other toxic metals in children living in areas close to mine tailings in southern Mexico, it was found that urinary Hg was elevated (Moreno et al. 2010). The individuals of interest were 50 children whose age ranged from 6 to 11 years; the results indicated that 30% of children had Hg levels above the reference value (0.7 μg L^{-1}) for urine.

In research intended to explore other routes of Hg exposure, Peregrino et al. (2011) analyzed the Hg content in skin-lightening creams manufactured in Mexico. It was noticed that none of the labels of the analyzed products gave Hg as being among the ingredients. In six of the creams, Hg concentrations were extremely elevated (from 878 to 36,000 ppm). Such Hg values imply a serious health risk, and indeed more research is needed on this topic.

The main functions of the human liver are to metabolize, transform, and store a wide variety of substances (Nieminen and Lemasters 1996). Moreover, considering its structure, function, and biochemistry, the liver is vulnerable to damage from excessive amounts of many toxic compounds (Timbrell 2009). Regarding effects on humoral immune responses, some investigations reported increased blastogenesis in human and animal lymphocytes when exposed in vitro to mercury (Exon et al. 1996; Rice 2001).

In Mexico, as in other developing countries, there is a need to study the toxicological effect of mercury and others pollutants (Yáñez et al. 2002). The main research areas that need attention as regards the effects of Hg on humans include (a) patterns of human consumption of predator fish species and Hg levels in the edible portion of these fish; (b) levels of Hg in people occupationally exposed to this element (personnel who formulate dental amalgams, miners, workers in chloralkali plants, etc.); and (c) concentration of Hg in skin-lightening creams manufactured in Mexico.

7 Summary

In Mexico, published studies relating to the occurrence of Hg in the environment are limited. Among the main sources of Hg in Mexico are mining and refining of Au and Hg, chloralkali plants, Cu smelting, residential combustion of wood, carboelectric plants, and oil refineries. Hg levels are highly variable in the atmospheric compartment because of the atmospheric dynamics and ongoing metal exchange with the terrestrial surface. In atmospheric studies, Hg levels are usually reported as total gaseous Hg (TGM). In Mexico, TGM values ranged from 1.32 ng m^{-3} in Hidalgo state (a rural agricultural area) to 71.82 ng m^{-3} in Zacatecas state (an area where brick manufacturers use mining wastes as a raw material).

Published information on mercury levels in the coastal environment comprise 21 studies, representing 21 areas, in which sediments constituted the substrate that was analyzed for Hg. In addition, water samples were analyzed for Hg in nine studies. Few studies exist on Hg levels in the Caribbean and in the southwest of the country where tourism is rapidly increasing. Hence, there is a need for establishing baseline levels of mercury in these increasingly visited areas. In regions where studies have been undertaken, Hg levels in sediments were highly variable. Variations in Hg sediment levels mainly result from geological factors and the varying degree of anthropogenic impacts in the studied areas. In areas that still have pristine or nearly pristine environments (e.g., coast, Baja California, Todos Santos Bay, and La Paz lagoon), sediment Hg levels ranged from <0.006 to 0.35 μg g^{-1} on a dry wt basis.

When higher levels exist (0.34–57.94 μg g^{-1} on a dry wt basis), the environment generally shows the influence of inputs from mining, oil processing, agriculture, geothermal events, or harmful algal bloom events (e.g., Guaymas Bay and Coatzacoalcos estuary). From chronological studies performed in selected coastal lagoons in NW Mexico, it is clear that Hg fluxes to sediments have increased from 2- to 15-fold in recent years. Since the 1940s, historical increases of Hg fluxes have resulted from higher agricultural waste releases and exhaust from the thermoelectric plants.

The levels of Hg in water reveal a moderate to elevated contamination of some Mexican coastal sites. In Urías lagoon (NW Mexico), moderate to high levels were found in the dissolved and suspended fraction, and these are related to shipping, the fishing industry, domestic effluents, and the presence of a thermoelectric plant. In Coatzacoalcos (SE Mexico), extremely elevated Hg levels were found during the decade of the 1970s. Low to moderate levels of Hg were measured in waters from the Alvarado lagoon (SE Mexico); those concentrations appear to be associated with river waters that became enriched with organic matter and suspended solids in the brackish mixing zone.

Regarding the Hg content in invertebrates, the use of bivalves (oysters and mussels) as biomonitors must be established along the coastal zones of Mexico, because some coastal lagoons have not been previously monitored. In addition, more research is needed to investigate shrimp farms that are associated with agricultural basins and receive effluents from several anthropogenic sources (e.g., mining activity and urban discharges).

Hg residues in several vertebrate groups collected in Mexico have been studied. These include mammals, birds, reptiles, and fish. In elasmobranch species, the highest Hg concentration (27.2 μg g^{-1} dry wt) was found in the muscle of the smooth hammerhead shark (*Sphyrna zygaena*). Teleost fish are the vertebrate group that has been most studied, with regard to Hg residue content; the highest value (5.67 μg g^{-1} dry wt) was detected in the striped marlin (*T. audax*). Among reptiles, only marine turtles were studied; Hg levels found ranged from 0.795 in the liver to 0.0006 μg g^{-1} dry wt in the blood of *L. olivacea*. In birds, the highest Hg concentration (5.08 μg g^{-1} dry wt) detected was in the liver of the olivaceous cormorant (*P. olivaceous*). Specimens from stranded marine mammals were also analyzed; levels of Hg ranged from 70.35 μg g^{-1} dry wt in the liver of stranded spinner dolphin (*S. longirostris*), to 0.145 μg g^{-1} dry wt in the muscle of gray whale (*E. robustus*). The presence of Hg in these marine animals is not thought to have caused the stranding of the animals.

Other organisms like macroalgae and vestimentiferan tube worms were used to monitor the occurrence of Hg in the aquatic environment; levels were comparable to data reported on similar organisms from other areas of the world. Few investigations have been carried out concerning the mercury content in human organs/tissues in Mexico. Considering the potential deleterious effects of Hg on kidney, lung, and the central nervous system, more information about human exposure to organic and inorganic forms of mercury and their effects is needed, both in Mexico and elsewhere.

Acknowledgements We acknowledge bibliographic support from Ms. C. Ramírez-Jáuregui and partial funding from the Ministry of Public Education (Project REDES PROMEP/103.5/12/4812).

References

Acosta-Saavedra LC, Moreno ME, Rodríguez-Kessler T, Luna A, Arias-Salvatierra D, Gómez R, Calderon-Aranda ES (2011) Environmental exposure to lead and mercury in Mexican children: a real health problem. Toxicol Mech Methods 21(9):656–666

Adams W, Kimerle RA, Barnett JW (1992) Sediment quality and aquatic life assessment. Environ Sci Technol 26:1874–1885

Al-Saleh I, Al-Doush I (2002) Mercury content in shrimp and fish species from the Gulf coast of Saudi Arabia. Bull Environ Contam Toxicol 68:576–583

Acosta y Asociados (2001) Inventario preliminar de emisiones atmosféricas de mercurio en México. Informe Final, México, pp 1–43

Aspmo K, Temme C, Berg T, Ferrari C, Gauchard PA, Fain X, Wibetoe G (2006) Mercury in the atmosphere, snow and melt water ponds in the North Atlantic Ocean during Arctic summer. Environ Sci Technol 40:4083–4089

Báez AP, Rosas I, Nulman R, Galvez L (1975) Movimiento del mercurio residual en el estuario del río Coatzacoalcos. Anal Inst Geofís (UNAM) 18:131–147

Botello AV, Páez-Osuna F (1986) El Problema Crucial: La Contaminación. Centro de Ecodesarrollo, Serie Medio Ambiente en Coatzacoalcos, Mexico, pp 1–180

Buchman M (1999) NOAA screening quick reference tables, Report 99-1, Coastal Protection and Restoration Division NOAA, Seattle, WA, pp 1–12

Buchman M (2008) NOAA screening quick reference tables, Report 08-1. Office of Response and Restoriation Division NOAA, Seattle, WA, pp 1–34

Cai Y, Rooker JR, Gill GA, Turner JP (2007) Bioaccumulation of mercury in pelagic fishes from the northern Gulf of Mexico. Can J Fish Aquat Sci 64:458–469

Carrasco L, Díez S, Soto D, Catalan J, Bayona J (2008) Assessment of mercury and methylmercury pollution with zebra mussel (*Dreissena polymorpha*) in the Ebro River (NE Spain) impacted by industrial hazardous dumps. Sci Total Environ 407:178–184

Carreón-Martínez LB, Huerta-Díaz MA, Nava-López C, Sequeiros-Valencia A (2001) Mercury and silver concentrations in sediments from the Port of Ensenada, Baja California, Mexico. Mar Pollut Bull 42:415–418

Carreón-Martínez LB, Huerta-Díaz MA, Nava-López C, Sequeiros-Valencia A (2002) Levels of reactive mercury and silver in sediments from the Port of Ensenada, Baja California, Mexico. Bull Environ Contam Toxicol 68:138–147

Chopin EIB, Alloway BJ (2007) Distribution and mobility of trace elements in soils and vegetation around the mining and smelting areas of Tharsis, Riotinto and Huelva, Iberian pyrite belt, SW Spain. Wat Air Soil Pollut 182:245–261

Chouvelon T, Warnau M, Churlaud C, Bustamante P (2009) Hg concentrations and related risk assessment in coral reef crustaceans, molluscs and fish from New Caledonia. Environ Pollut 157:331–340

Cohen JT, Bellinger DC, Connor WE, Kris-Etherton PM, Lawrence RS, Savitz DA (2005) A quantitative risk-benefit analysis of change in population fish consumption. Am J Prev Med 29: 325–334

CONABIO (2004) Comisión Nacional para el Conocimiento Uso de la Biodiversidad. Regionalización. http://www.conabio.gob.mx

Cosson RP (1996) La bioaccumulation des éléments minéraux chez le vestimentifère *Riftia pachyptila* (Jones): bilan de connaissances. Oceanol Acta 2:163–176

De la Rosa DA, Volke-Sepúlveda T, Solórzano G, Green C, Tordon R, Beauchamp S (2004) Survey of atmospheric total geaseous mercury in Mexico. Atmos Environ 38:4839–4846

De-la-Peña-Sobarzo P (2003) Focos rojos de mercurio en América del Norte. El Faro 31:8–9

Ebinghaus R, Jennings SG, Kock HH, Derwent RG, Manning AJ, Spain TG (2011) Decreasing trends in total gaseous mercury observations in baseline air at Mace Head, Ireland from 1996 to 2009. Atmos Environ 45:3475–3480

Echeverria D, Woods JS, Heyer NJ, Rohlman DS, Farin FM, Bittner AC Jr, Li T, Garabedian C (2005) Chronic low-level mercury exposure, BDNF polymorphism, and associations with cognitive and motor function. Neurotoxicol Teratol 27:781–796

Eisler R (1987) Mercury hazards to fish, wildlife, and invertebrates: a synoptic review. Biological Report (No. 1.1), U.S. Fish and Wildlife Service, USA, pp 1–85

Eisler R (2010) Compendium of trace metals and marine biota. Volume 2, Vertebrates, 1st edn. Elsevier, Amsterdam, pp 110–140

Elahi M, Esmaili-Sari A, Bahramifar N (2007) Total mercury levels in selected tissues of some marine crustaceans from Persian Gulf, Iran: variations related to length, weight and sex. Bull Environ Contam Toxicol 88:60–64

Escobar-Sánchez O, Galván-Magaña F, Rosíles-Martínez R (2010) Mercury and selenium bioaccumulation in the smooth hammerhead shark, *Sphyrna zygaena* Linnaeus, from the Mexican Pacific Ocean. Bull Environ Contam Toxicol 84:488–491

Escobar-Sánchez O, Galván-Magaña F, Rosiles-Martínez R (2011) Biomagnification of mercury and selenium in blue shark *Prionace glauca* from the Pacific ocean off Mexico. Biol Trace Elem Res 144(1–3):550–559

Exon JH, South EH, Hendrix K (1996) Effects of metals on the humoral immune response. In: Chang LW (ed) Toxicology of metals. Lewis Publishers, Boca Raton, FL, pp 1106–1198

Fergusson JE (1990) The heavy elements, chemistry, environmental impact and health effects. Pergamon Press, Oxford, pp 1–614

Fitzgerald WF, Lamborg CH (2005) Geochemistry of mercury in the environment. In: Holland HD, Turekian KK (eds) Treatise on geochemistry, environmental geochemistry. Elsevier, San Diego, USA, pp 108–148

Fitzgerald WF, Mason RP (1996) The global mercury cycle: oceanic and anthropogenic aspects. In: Baeyens W et al (eds) Global and regional mercury cycles: sources, fluxes and mass balances. Kluwer Academic Publishers, Dordrecht, pp 85–108

Fowler BA (1996) The nephropathology of metals. In: Chang CH (ed) Toxicology of metals. Lewis Publishers, Boca Raton, FL, pp 1–1198

Friberg L, Monet NK (1989) Accumulation of methylmercury and inorganic mercury in the brain. Biol Trace Elem Res 21:201–206

Fu XW, Feng X, Dong ZQ, Yin RS, Wang JX, Yang ZR, Zhang H (2010) Atmospheric gaseous elemental mercury (GEM) concentrations and mercury depositions at a high-altitude mountain peak in south China. Atmos Chem Phys 10:2425–2437

García-Hernández J, García-Rico L, Jara-Marini ME, Barraza-Guardado R, Hudson WA (2005) Concentrations of heavy metals in sediment and organisms during a harmful algal bloom (HAB) at Kun Kaak Bay, Sonora, Mexico. Mar Pollut Bull 50:733–739

García-Hernández J, Cadena-Cárdenas L, Betancourt-Lozano M, García-de-la-Parra LM, García-Rico L, Márquez-Farías F (2007) Total mercury content found in edible tissues of top predator fish from the Gulf of California, Mexico. Toxicol Environ Chem 89(3):507–522

García-Rico L, Wilson-Cruz S, Frasquillo-Felix MC, Jara-Marini ME (2003) Total metals in intertidal surface sediment of oyster culture areas in Sonora, Mexico. Bull Environ Contam Toxicol 70:1235–1241

García-Rico L, Valenzuela-Rodríguez M, Jara-Marini ME (2006) Geochemistry of mercury in sediments of oyster areas in Sonora, Mexico. Mar Pollut Bull 52:453–458

García-Rico L, Tejeda-Valenzuela L, Burgos-Hernández A (2010) Seasonal variations in the concentrations of metals in *Crassostrea corteziensis* from Sonora, Mexico. Bull Environ Contam Toxicol 85:209–213

Gaskin DE, Smith GJD, Arnold PW, Lousy MV, Frank R, Holdrinet M, Wade MC (1974) Mercury, DDT, dieldrin and PCB in two species of Odontoceti (Cetacea) from St Lucia, Lesser Antilles. J Fish Res Board Can 31:1235–1239

Gonzalez-Ramírez D, Maiorino RM, Zuñiga-Charles M, Xu Z, Hurlbut KM, Junco-Muñoz P, Aposhian MM, Dart RC, Diaz-Gama JH, Echeverria D (1995) Sodium 2,3-dimercaptopropane-1-sulfonate challenge test for mercury in humans. J Pharmacol Exp Ther 272:264–274

Green-Ruiz C, Ruelas-Inzunza J, Páez-Osuna F (2005) Mercury in surface sediments and benthic organisms from Guaymas Bay, east coast of the Gulf of California. Environ Geochem Health 27:321–329

Guentzel JL, Portilla E, Keith KM, Keith EO (2007) Mercury transport and bioaccumulation in riverbank communities of the Alvarado lagoon system, Veracruz, state, Mexico. Sci Total Environ 388:316–324

Gustin MS, Lindberg SE, Weisberg PJ (2008) An update on the natural sources and sinks of atmospheric mercury. Appl Geochem 23:482–493

Gutiérrez-Galindo EA, Flores-Muñoz G, Aguilar-Flores A (1988) Mercury in freshwater fish and clams from the Cerro Prieto geothernal field of Baja California, Mexico. Bull Environ Contam Toxicol 41:201–207

Gutiérrez-Galindo EA, Casas-Beltrán DA, Muñoz-Barbosa A, Macías-Zamora JV, Segovia-Zavala JA, Orozco-Beltrán MV, Daessle LW (2007) Spatial distribution and enrichment of mercury in surface sediments off the northwest coast of Baja California, Mexico. Cienc Mar 33:473–482

Gutiérrez-Galindo EA, Casas-Beltrán DA, Muñoz-Barbosa A, Daessle LW, Segovia-Zavala JA, Macías-Zamora JV, Orozco-Beltrán MV (2008) Distribution of mercury in surficial sediments from Todos Santos Bay, Baja California, Mexico. Bull Environ Contam Toxicol 80:123–127

Gutiérrez-Mejía E, Lares ML, Sosa-Nishizaki O (2009) Mercury and arsenic in muscle and liver of the golden cownose ray, *Rhinoptera steindachneri*, Evermann and Jenkins, 1891, from the upper Gulf of California, México. Bull Environ Contam Toxicol 83:230–234

Henry CD (1989) Late Cenozoic basin and range structure in western Mexico adjacent to the Gulf of California. Geol Soc Am Bull 101:1147–1155

Huerta-Díaz MA, Morse JW (1990) A quantitative method for determination of trace metal concentrations in sedimentary pyrite. Mar Chem 29:119–144

Hui CA, Takekawa JY, Warnock SE (2001) Contaminant profiles of two species of shorebirds foraging together at two neighboring sites in south San Francisco Bay, California. Environ Monitor Assess 71:107–121

Hurtado-Banda R, Gomez-Alvarez A, Márquez-Farías JF, Cordoba-Figueroa M, Navarro-García G, Medina-Juárez LA (2012) Total mercury in liver and muscle tissue of two coastal sharks from the northwest of Mexico. Bull Environ Contam Toxicol 88(6):971–975

Jara Marini M, Soto-Jiménez MF, Páez-Osuna F (2008a) Bulk and bioavailable heavy metals (Cd, Cu, Pb and Zn) in surface sediments from Mazatlán Harbor (SE Gulf of California). Bull Environ Contam Toxicol 80(2):150–153

Jara Marini M, Soto-Jiménez MF, Páez-Osuna F (2008b) Trace metal accumulation patterns in a mangrove lagoon ecosystem, Mazatlán Harbor, southeastern Gulf of California. J Environ Sci Health A Tox Hazard Subst Environ Eng 43:995–1005

Kampalath R, Gardner SC, Mendez-Rodriguez L, Jay JA (2006) Total and methylmercury in three species of sea turtles of Baja California Sur. Mar Pollut Bull 52:1816–1823

Kang H, Xie Z (2011) Atmospheric mercury over the marine boundary layer observed during the third China Arctic Research Expedition. J Environ Sci 23(9):1424–1430

Kot FS, Green-Ruiz C, Páez-Osuna F, Shumilin E, Rodriguez-Meza D (1999) Distribution of mercury in sediments from La Paz lagoon, peninsula of Baja California, Mexico. Bull Environ Contam Toxicol 63:45–51

Kot F, Shumilin E, Rodriguez-Figueroa GM, Mirlean N (2009) Mercury dispersal to Arroyo and coastal sediments from abandoned copper mine operations, El Boleo, Baja California. Bull Environ Contam Toxicol 82:20–25

Kwokal Z, Bilinski FS, Bilinski H, Branica M (2002) A comparison of anthropogenic mercury pollution in Kastela Bay (Croatia) with pristine estuaries in Ore (Sweden) and Krka (Croatia). Mar Pollut Bull 44:1152–1169

Landis WG, Yu MH (1999) Introduction to environmental toxicology. Lewis Publishers, Boca Raton, FL, pp 1–390

Lawrence B (2000) The mercury marketplace: sources, demand, price, and the impacts of environmental regulation. Presentation at USEPA's Workshop on mercury in products, processes, waste, and the environment. Baltimore, MD, 22–23 March 2000, as quoted by USA (comm-24-gov)

Leal MCF, Vasconcelos MT, Sousa-Pinto I, Cabral JPS (1997) Biomonitoring with benthic macroalgae and direct assay of heavy metals in seawater of the Oporto coast (Northwest Portugal). Mar Pollut Bull 34(12):1006–1015

Leal-Acosta ML, Shumilin E, Mirlean N, Sapozhnikov D, Gordeev V (2010) Arsenic and mercury contamination of sediments of geothermal springs, mangrove lagoon and the Santispac Bight, Bahía Concepción, Baja California peninsula. Bull Environ Contam Toxicol 85:609–613

Levy M, Schwartz S, Dijak M, Weber JP, Tardif R, Rouah F (2004) Childhood urine mercury excretion: dental amalgam and fish consumption as exposure factors. Environ Res 94: 283–290

Ley-Quiñónez C, Zavala-Norzagaray AA, Espinosa-Carreón TL, Peckham H, Marquez-Herrera C, Campos-Villegas L, Aguirre AA (2011) Baseline heavy metals and metalloids values in blood of loggerhead turtles (*Caretta caretta*) from Baja California Sur, Mexico. Mar Pollut Bull 62:1979–1983

Li P, Feng X, Qiu G, Li Z, Fu X, Sakamoto M, Liu X, Wang D (2008) Mercury exposures and symptoms in smelting workers of artisanal mercury mines in Wuchuan, Guizhou, China. Environ Res 107:108–114

Li P, Feng X, Qiu G, Shang L, Wang S (2011) Mercury pollution in Wuchuan mercury mining area, Guizhou, Southwestern China: the impacts from large scale and artisanal mercury mining. Environ Int 42:59–66

Long ER, MacDonald DD, Smith SL, Calder FD (1995) Incidente of adverse biological effects within ranges of chemical concentrations in marine and estuarine sediments. Environ Manage 19:18–97

López M, González I, Romero A (2008) Trace element contamination of agricultural soils affected by sulphide exploitation (Iberian Pyrite Belt, SW Spain). Environ Geol 54:805–818

Mackay NJ, Kazacos MN, Williams RJ, Ledow ML (1975) Selenium and heavy metals in black marlin. Mar Pollut Bull 6:57–60

Manisseri MK, Menon MN (1995) Copper-induce damage to hepatopancreas of the penaeid shrimp *Metapenaeus dobsoni* an ultrastructural study. Dis Aquat Organ 22:51–57

Markert BA, Breure AM, Zechmeister HG (2003) Definitions, strategies and principles for bioindication/biomonitoring of the environment. In: Markert BA, Breure AM, Zechmeister HG (eds) Bioindicators & biomonitors. Principles, concepts and applications. Elsevier, Amsterdam, pp 3–40

Mason RP, Fitzgerald WF, Morel FMM (1994) The biogeochemical cycling of elemental mercury: anthropogenic influences. Geochim Cosmochim Acta 58:3191–3198

Maz-Courrau A, López-Vera C, Galván-Magaña F, Escobar-Sánchez O, Rosíles-Martínez R, Sanjuán-Muñoz A (2012) Bioaccumulation and biomagnification of total mercury in four exploited shark species in the Baja California Peninsula, Mexico. Bull Environ Contam Toxicol 88(2):129–134

McDonald DD, Carr RS, Calder FD, Long ER, Ingersoll CG (1996) Development and evaluation of sediment quality guidelines for Florida coastal waters. Ecotoxicology 5:253–278

Moreno ME, Acosta-Saavedra LC, Meza-Figueroa D, Vera E, Cebrian ME, Ostrosky-Wegman P, Calderon-Aranda ES (2010) Biomonitoring of metal in children living in a mine tailings zone in southern Mexico: a pilot study. Int J Hyg Environ Health 213:252–258

Nieminen AL, Lemasters JJ (1996) Hepatic injury by metal accumulation. In: Chang LW (ed) Toxicology of metals. Lewis Publishers, Boca Raton, FL, pp 1–1198

Nriagu JO (1989) A global assessment of natural sources of atmospheric trace metals. Nature 338:47–49

Nriagu JO, Pacyna JM (1988) Quantitative assessment of worldwide contamination of air, water and soils by trace metals. Nature 333:134–139

Núñez-Nogueira G (2005) Concentration of essential and non-essential metals in two shark species commonly caught in Mexican (Gulf of Mexico) coastline. In: Botello AV, Rendón-von Osten J, Gold-Bouchot G, Agraz-Hernández C (eds) Golfo de México Contaminación e Impacto Ambiental: Diagnóstico y Tendencias. Universidad Autónoma de Campeche, Universidad Nacional Autónoma de México, Instituto Nacional de Ecología, México, pp 451–474

Núñez-Nogueira G, Bautista-Ordóñez J, Rosiles-Martínez R (1998) Concentración y distribución de mercurio en tejidos del cazón (*Rhizoprionodon terraenovae*). Vet Mexico 29(1):15–21

O'Connor TP, Daskalakis KD, Hyland JL, Paul JF, Summers JK (1998) Comparisons of sediment toxicity with predictions based on chemical guidelines. Environ Toxicol Chem 17:468–471

Ochoa SA, Halffter GE, Ibarra, R (1973) Estudio de la Contaminación en el Bajo río Coatzacoalcos. In: Primer Seminario sobre la Evaluación de la Contaminación Ambiental. Instituto Mexicano de Recursos Naturales. Escuela Nacional de Ciencias Biológicas, Instituto Politécnico Nacional, México, pp 115–162

Ordiano-Flores A, Galván-Magaña F, Rosiles-Martínez R (2011) Bioaccumulation of mercury in muscle tissue of yellowfin tuna, *Thunnus albacares*, of the Eastern Pacific Ocean. Biol Trace Elem Res 144:606–620

Osuna-Martínez CC, Páez-Osuna F, Alonso-Rodríguez R (2010) Mercury in cultured oysters (*Crassostrea gigas* Thunberg, 1793 and *C. corteziensis* Hertlein, 1951) from four coastal lagoons of the SE Gulf of California, Mexico. Bull Environ Contam Toxicol 85:339–343

Páez-Osuna F, Ramírez-Reséndiz G, Ruiz Fernández AC, Soto-Jiménez MF (2007) La Contaminación por Nitrógeno y Fósforo en Sinaloa: Flujos, Fuentes, Efectos y Opciones de manejo. In: Páez-Osuna F (ed) Serie Lagunas Costeras de Sinaloa. UNAM, El Colegio de Sinaloa, México, pp 1–304

Páez-Osuna F, Calderón-Campuzano MF, Soto-Jiménez MF, Ruelas-Inzunza J (2011) Mercury in blood and eggs of the sea turtle *Lepidochelys olivacea* from a nesting colony in Oaxaca, Mexico. Mar Pollut Bull 62:1320–1323

Peregrino CP, Moreno MV, Miranda SV, Rubio AD, Leal LO (2011) Mercury levels in locally manufactured Mexican skin-lightening creams. Int J Environ Res Public Health 8:2516–2523

Pérez-Zapata AJ (1981) Plomo y mercurio. In: Centro de Ecosdesarrollo (ed) Lagunas Costeras de Tabasco. Un ecosistema en peligro. Centro de Ecosdesarrollo, México, D.F, pp 58–61

Pérez-Zapata AJ, Peleón IR, Gil RAM (1984) Determinación cuantitativa de plomo en peces del estuario del Río Coatzacoalcos. Anal Esc Nac Cienc Biol México 28:193–197

Phillips DJH, Ho CT, Ng LH (1982) Trace elements in the Pacific oyster in Hong Kong. Arch Environ Contam Toxicol 11:533–537

Pirrone N, Keeler GJ, Nriagu JO (1996) Regional differences in worldwide emissions of mercury to the atmosphere. Atmos Environ 30(37):2981–2987

PNUMA (Programa de las Naciones Unidas para el Medio Ambiente) (2005) Evaluación mundial sobre el mercurio, Ginebra, pp 1–289

Presti SM, Resendiz-Hidalgo A, Sollod AE, Seminoff JA (1999) Mercury concentration in the scutes of Black Sea turtles, *Chelonia mydas agassizii*, in the Gulf of California. Chelonian Conserv Biol 3(3):531–533

Rahman SA, Wood AK, Sarmani S, Majid AA (1997) Determination of mercury and organic mercury contents in Malaysian seafood. J Radioanal Nucl Chem 217:53–56

Reimer AA, Reimer RD (1975) Total mercury in some fish and shellfish along the Mexican coast. Bull Environ Contam Toxicol 14(1):105–111

Rice C (2001) Fish immunotoxicology: understanding mechanisms of action. In: Schlenk D, Benson WH (eds) Target organ toxicity in marine and freshwater teleosts. Taylor & Francis, London, pp 97–140

Riisgard HG, Famme P (1986) Accumulation of inorganic and organic mercury in shrimp, *Crangon crangon*. Mar Pollut Bull 17:255–257

Roja de Astudillo L, Chang-Ye I, Agard J, Bekele I, Hubbard R (2002) Heavy metals in green mussel (*Perna viridis*) and oysters (*Crassostrea* sp.) from Trinidad and Venezuela. Arch Environ Contam Toxicol 42:410–415

Romero-Vargas IP (1995) Metales pesados y su fraccionación química en la Bahía de Todos Santos, Baja California, México. B Sc Thesis, Universidad Autónoma de Baja California, Ensenada, B.C., pp 1–86

Rosas I, Báez A, Belmont R (1983) Oyster (*Crassostrea virginica*) as indicator of heavy metal pollution in some lagoons of the Gulf of Mexico. Wat Air Soil Pollut 20:127–135

Ruelas-Inzunza J, Páez-Osuna F (2005) Mercury in fish and shark tissues from two coastal lagoons in the Gulf of California, Mexico. Bull Environ Contam Toxicol 74:294–300

Ruelas-Inzunza J, Páez-Osuna F, Pérez-Cortés H (2000) Distribution of mercury in muscle, liver and kidney of the spinner dolphin (*Stenella longirostris*) stranded in the southern Gulf of California. Mar Pollut Bull 40(11):1063–1066

Ruelas-Inzunza J, Horvat M, Pérez-Cortés H, Páez-Osuna F (2003a) Methylmercury and total mercury distribution in tissues of gray whales (*Eschrichtius robustus*) and spinner dolphins (*Stenella longirostris*) stranded along the lower Gulf of California, Mexico. Cienc Mar 29(1):1–8

Ruelas-Inzunza J, Soto LA, Páez-Osuna F (2003b) Heavy-metal accumulation in the hydrothermal vent clam *Vesicomya gigas* from Guaymas basin, Gulf of California. Deep Sea Res (Part I) 50:757–761

Ruelas-Inzunza J, García-Rosales SB, Páez-Osuna F (2004) Distribution of mercury in adult penaeid shrimp from Altata-Ensenada del Pabellón lagoon (SE Gulf of California). Chemosphere 57:1657–1661

Ruelas-Inzunza J, Páez-Osuna F, Soto LA (2005) Bioaccumulation of Cd, Co, Cr, Cu, Fe, Hg, Mn, Ni, Pb and Zn in trophosome and vestimentum of the tube worm *Riftia pachyptila* from Guaymas basin, Gulf of California. Deep Sea Res (Part I) 52:1319–1323

Ruelas-Inzunza J, Páez-Osuna F, Arvizu-Merin M (2007) Mercury distribution in selected tissues of migratory and resident avifauna from Altata-Ensenada del Pabellón lagoon, Southeast Gulf of California. Bull Environ Contam Toxicol 78(1):39–43

Ruelas-Inzunza J, Meza-López G, Páez-Osuna F (2008) Mercury in fish that are of dietary importance from the coasts of Sinaloa (SE Gulf of California). J Food Compost Anal 21: 211–218

Ruelas-Inzunza J, Páez-Osuna F, Zamora-Arellano N, Amezcua-Martínez F, Bojórquez-Leyva H (2009) Mercury in biota and surficial sediments from Coatzacoalcos estuary, Gulf of Mexico: distribution and seasonal variation. Wat Air Soil Pollut 197:165–174

Ruelas-Inzunza JR, Páez-Osuna F, Ruiz-Fernández AC, Zamora-Arellano N (2011a) Health risk associated to dietary intake of mercury in selected coastal areas of Mexico. Bull Environ Contam Toxicol 86:180–188

Ruelas-Inzunza J, Patiño-Mejía C, Soto-Jiménez M, Barba-Quintero G, Spanopoulos-Hernández M (2011b) Total Hg in canned yellowfin tuna *Thunnus albacares* marketed in northwest Mexico. Food Chem Toxicol 49:3070–3073

Ruiz-Fernández AC, Frignani M, Hillaire-Marcel C, Ghaleb B, Arvizu MD, Raygoza-Viera JR, Páez-Osuna F (2009) Trace metals (Cd, Cu, Hg and Pb) accumulation recorded in the intertidal mudflat sediments of three coastal lagoons of the Gulf of California, México. Estuaries Coasts 32(3):551–560

Rutter AP, Snyder DC, Stone EA, Schauer JJ, Gonzalez-Abraham R, Molina LT, Márquez C, Cárdenas B, de-Foy B (2009) In situ measurements of speciated atmospheric mercury and the identification of source regions in the Mexico city metropolitan area. Atmos Chem Phys 9:207–220

Rytuba JJ (2003) Mercury from mineral deposits and potential environmental impact. Environ Geol 43:326–338

Sadiq M (1992) Toxic metal chemistry in marine environments. Marcel Dekker Inc, New York, pp 1–390

Soto-Jiménez MF, Amezcua F, González-Ledesma R (2010) Nonessential metals in striped marlin and Indo-Pacific sailfish in the Southeast Gulf of California, Mexico: concentration and assessment of human health risk. Arch Environ Contam Toxicol 58:810–818

Storelli MM, Ceci E, Storelli A, Marcotrigiano GO (2003) Polychlorinated biphenyl, heavy metal, and methylmercury residues in hammerhead sharks: contaminant status and assessment. Mar Pollut Bull 46:1035–1048

Timbrell JA (2009) Principles of biochemical toxicology. Informa, London, pp 1–453

Trasande L, Cortes JF, Landrigan PJ, Abercrombie MI, Bopp RF, Cifuentes E (2010) Methylmercury exposure in a subsistence fishing community in Lake Chapala, Mexico: an ecological approach. Environ Health 9:1–10

Turekian AV, Wedepohl KH (1961) Distribution of the elements in some major units of the earth's crust. Bull Geol Soc Am 72:175–192

US EPA (1997) Mercury study report to congress. http://www.epa.gov/mercury/report.htm

US EPA (2006) National recommended water quality criteria. http://www.epa.gov/waterscience/criteria/nrwqc-.pdf

Vaisman AG, Marins RV, Lacerda LD (2005) Characterization of the mangrove oyster, Crassostrea rhizophorae, as a biomonitor for mercury in tropical estuarine systems, Northeast Brazil. Bull Environ Contam Toxicol 74:582–588

Valente RJ, Shea C, Humes KL, Tanner RL (2007) Atmospheric mercury in the Great Smoky Mountains compared to regional and global levels. Atmos Environ 41:1861–1873

Varanasi U, Stein JE, Tilbury KL, Meador JP, Sloan CA, Clark RC, Chan SL (1994) Chemical contaminants in gray whales (*Eschrichtius robustus*) stranded along the west coast of North America. Sci Total Environ 145:29–53

Vázquez F, Florville-Alejandre TR, Herrera M, Díaz-de-León LM (2008) Heavy metals in muscular tissue of the catfish, *Ariopsis felis*, in the southern Gulf of Mexico (2001–2004). Lat Am J Aquat Res 36(2):223–233

Velasco-González OH, Echavarría-Almeida S, Pérez-López ME, Villanueva-Fierro I (2001) Contenido de mercurio y arsénico en atún y sardinas enlatadas mexicanas. Rev Int Contam Ambient 17(1):31–35

Villanueva SF, Botello AV (1998) Metal pollution in coastal areas of Mexico. Rev Environ Contam Toxicol 157:53–94

Wang X, Sato T, Xing B, Tao S (2002) Health risk of heavy metals to the general public in Tianjin, China via consumption of vegetables and fish. Sci Total Environ 350:28–37

White RF, Feldman RG, Proctor SP (1992) Neurobehavioral effect of toxic exposures. In: White RF (ed) Clinical syndromes in adult neuropsicology. Elsevier, Amsterdam, pp 1–512

Willerer AOM, Kot FS, Shumilin EN, Lutsarev S, Rodriguez AJM (2003) Mercury in bottom sediments of the Tropical Rio Marabasco, its estuary, and Laguna de Navidad, Mexico. Bull Environ Contam Toxicol 70:1213–1219

Woods JS, Martin MD, Leroux BG, DeRouen TA, Leitao JG, Bernardo MF, Luis HS, Simmonds PL, Kushleika JV, Huang Y (2007) The contribution of dental amalgam to urinary mercury excretion in children. Environ Health Perspect 115:1527–1531

World Health Organization (WHO) (1990) International program on chemical safety. Environmental Health Criteria 101, Rome

Xue F, Holzman C, Rahbar MH, Trosko K, Fisher L (2007) Maternal fish consumption, mercury levels, and risk of preterm delivery. Environ Health Perspect 115:42–47

Yáñez L, Ortiz D, Calderón J, Batres L, Carrizales L, Mejía J, Martínez L, García-Nieto E, Díaz-Barriga F (2002) Overview of human health and chemical mixtures: problems facing developing countries. Environ Health Perspect 110:901–909

Zorita AI, Ortiz-Zarragoitia M, Orbea A, Cancio I, Soto M, Marigómez I, Cajaraville M (2007) Assessment of biological effects of environmental pollution along the NW Mediterranean Sea using mussels as sentinel organisms. Environ Pollut 148:236–250

Do Cd, Cu, Ni, Pb, and Zn Biomagnify in Aquatic Ecosystems?[*]

Rick D. Cardwell, David K. DeForest, Kevin V. Brix, and William J. Adams

Contents

---

[*]Additional material for this chapter can be found on http://extras.springer.com

R.D. Cardwell (✉)
Cardwell Consulting LLC, 2193 NW Kinderman Place, Corvallis, OR 97330, USA
e-mail: rcardwell6@gmail.com

D.K. DeForest
Windward Environmental, 200 West Mercer St., Suite 401, Seattle, WA 98119, USA
e-mail: DavidD@windwardenv.com

K.V. Brix
EcoTox, 575 Crandon Blvd., #703, Key Biscayne, FL 33149, USA

RSMAS, University of Miami, 4600 Rickenbacker Causeway, Miami, FL 33149, USA
e-mail: kbrix@rsmas.miami.edu

W.J. Adams
Rio Tinto, 4700 Daybreak Parkway, South Jordan, UT 84095, USA
e-mail: William.Adams@riotinto.com

D.M. Whitacre (ed.), *Reviews of Environmental Contamination and Toxicology*
Volume 226, Reviews of Environmental Contamination and Toxicology 226,
DOI 10.1007/978-1-4614-6898-1_4, © Springer Science+Business Media New York 2013

1   Introduction

Trophic transfer (biotransference—Dallinger et al. 1987) results from passage of a contaminant through food chains as a result of uptake only from water (bioconcentration), only from diet (dietary accumulation), or from a combination of these (bioaccumulation) (Biddinger and Gloss 1984; Davis and Foster 1958; Macek et al. 1979; Suedel et al. 1994). Trophic transfer factors (TTFs) are analogous to bioaccumulation (accumulation) factors, the original terms used to describe steady-state tissue residues in an organism resulting from both water and dietary uptake pathways (Boroughs et al. 1957). TTFs are the same as biomagnification factors and also meet the definition of biomagnification when TTFs exceeding 1.0 are observed through three or more trophic levels as a result of at least two trophic transfers (Biddinger and Gloss 1984). Most investigators have assumed TTFs result mainly from dietary accumulation (Baptist and Lewis 1967; Mathews and Fisher 2008; Reinfelder et al. 1998), although it is not possible to distinguish aqueous from dietary uptake in field studies. Moreover, the relative importance of the diet and aqueous uptake pathways is context-dependent, varying with exposure duration, metal bioavailability, and the species and their prey. Application of the aquatic TTF concept, as currently understood, may have first been proposed by Baptist and Lewis (1967), and the term had been widely adopted by the early 1990s (Baudin and Nucho 1992; Dillon et al. 1995; Garnier-Laplace et al. 1997; Suedel et al. 1994).

All the foregoing processes—bioconcentration, bioaccumulation, biomagnification, and trophic transfer—were well described and defined by the 1950s, albeit with differences in specific usage. Most of these terms were originally introduced by scientists studying the environmental fate of radionuclides (e.g., Baptist and Lewis 1967; Davis and Foster 1958; Krumholz and Foster 1957). The toxicological implications of biomagnification were established decades ago for many compounds, such as mercury in the late 1950s (McAlpine and Araki 1959), DDT in the 1960s (Burdick et al. 1964; Hunt 1966; Peakall 1969), and PCBs (polychlorinated biphenyls) and chlorinated insecticides in the early 1970s (Hunt 1966; Peakall and Lincer 1970). However, such was not the case for most metals. Biomagnification has been shown to be a predictor of aquatic hazard for certain nonpolar hydrophobic compounds having specific properties, such as log K_{ow} >6.0, depuration rate half lives >40 days, and assimilation efficiencies >35% (Borga et al. 2011; Bruggeman et al. 1984; Fisk et al. 1998; Macek et al. 1979). Inorganic metals, in contrast, are hydrophilic, some are metabolically essential (Davis and Gatlin 1996; White and Rainbow 1985), and most are variably regulated and detoxified (Ju et al. 2011).

Trophic transfer and biomagnification historically were calculated similarly, but now are calculated in various ways. TTFs can be based on single or multiple transfers (steps) in a food chain, with each step representing a predator–prey interaction, such as algae → planktor or oyster → carnivorous snail. For many years, biomagnification typically was calculated the same way. Both of these methods differ from that used to calculate trophic magnification factors (Borga et al. 2011), which is the antilog of a linear regression slope relating trophic level to log tissue concentration in field-based food web studies. A third method for calculating trophic transfer uses a biokinetic model that measures the influences of exposure concentration, assimilation efficiency, and growth (Reinfelder et al. 1998; Luoma and Rainbow 2005). Because trophic transfer is chemical- and species-specific (Wang 2002), it follows that trophic transfer and biomagnification may be food chain-specific, hence varying between lab studies and field studies.

In this paper, we evaluate published data in which single and multiple trophic transfers of five metals (Cd, Cu, Pb, Ni, and Zn) were studied in freshwater and marine food chains. We examine whether the aquatic toxicity of bioaccumulated metals may be related to TTFs, and we compare TTFs measured in the laboratory and field with those estimated using biokinetic models. In performing this review, we had the following objectives: First, to compare TTFs generated from lab and field data on tissue residues in predator and prey. Second, to compare these TTFs to those estimated using a standard biokinetic model (see review of Reinfelder et al. 1998). Third, we sought to determine whether relationships existed between TTF magnitude and metal concentration in prey. Fourth, we evaluated relationships between TTF magnitude and toxicity, which is the key reason that biomagnification in aquatic food webs is of concern (Meador 2006). The final objective was to consider whether TTFs and tissue residues for metals that were required for metabolism (Cu, Ni, and Zn) differed from those that were not required (Cd and Pb) (Brown 2005).

2 Methods

Relevant studies from the peer-reviewed literature cogent to our topic were identified using online searches (e.g., Google Scholar, SCIRUS, and ISI Web of Knowledge). We focused on Cd, Cu, Pb, Ni, and Zn because these are relatively data rich and represent essential (Cu, Ni, Zn) and non-essential (Cd and Pb) elements.

2.1 Laboratory Data

Trophic transfer data were compiled from dietary accumulation studies conducted in the laboratory (Online Resource 1). In these studies, measurements of metal concentration in the diet were made.

We compiled TTF data from studies in which the metal was incorporated into a natural food or was spiked into a processed food, such as a pellet. These data reflected tests having varying exposure durations, diet types and rations, some of which do not occur in nature. These differences may be consequential because metal bioavailability and assimilation efficiency will vary between diet types from a variety of factors: differences in the form in which the metal is stored, species-specific differences in metal homeostatic mechanisms (Wallace and Lopez 1997; Wallace et al. 2003; Ju et al. 2011), and differences in metal exposure history

(Rainbow et al. 2006). Only data in which the metal was measured in the consumer and its food were used. The TTFs were based on whole body metal concentrations so that comparisons across species and trophic levels could be normalized to the most commonly measured tissue type. We also evaluated whether TTF magnitudes and toxicity were related where both were measured.

A typical laboratory study design was to expose algae to a metal in an aqueous medium and then to feed the algae to a planktivore. In lab studies with carnivores, prey often were exposed via water or diet or both, then fed to the predator either live or in a processed form (e.g., as freeze-dried pellets). Trophic transfer factors were compiled from studies in which both single (e.g., alga → herbivore) and multiple (e.g., alga → herbivore → carnivore) steps in food chains were evaluated. Although the latter data were more relevant for evaluating metal biomagnification potential, one-step TTF data facilitated evaluation of relationships between TTFs and exposure concentration as well as those between TTFs and toxicity. Variability in these relationships is expected when the exposure concentration of a given metal influences the kinetics of uptake and depuration, and ultimately, TTF magnitudes in a species (Reinfelder et al. 1998) and toxicity (Cheung et al. 2006; Croteau and Luoma 2009).

All dietary and whole body tissue concentrations were expressed as dry weights. In some cases, typically fish, residues had to be calculated from wet weight concentrations using measured or assumed (75%) moisture contents. The moisture contents of whole body fish appeared to be relatively consistent, with reported values of 71, 75, and 77% ($n = 2,051$), corresponding to 25th, 50th, and 75th quartiles, respectively (Seiler and Skorupa 2001).

Relationships between $\log_{10}$ TTFs and $\log_{10}$ dietary metal concentrations were plotted and the slopes tested for statistical significance ($a = 0.05$) using linear regression. A significant negative slope—an inverse relationship—indicates that the metal concentration tends to be lower in the consumer than in its food. A significant positive slope indicates that a metal's concentration tends to be greater in the consumer than in its food. Slopes may be positive or negative, reflecting inherent differences among species (Schmidt et al. 2011), and the properties of the environment from which the species were sampled (Borga et al. 2011). Moreover, one-step increases in a metal concentration neither meet the definition of biomagnification (encompassing at least three trophic levels), nor that of trophic magnification.

2.2 Laboratory Biokinetic Data

TTFs were also estimated from biokinetic models that were parameterized, using values from the literature. Steady-state metal concentrations from waterborne and dietary exposures were estimated using the following equation (Reinfelder et al. 1998):

$$C_{ss} = \frac{(k_u \times C_w) + (AE \times IR \times C_f)}{k_e + k_g}, \tag{1}$$

where, C_{ss} = steady-state metal concentration (μg g^{-1}); k_u = uptake rate constant (L g^{-1} day^{-1}); C_w = metal concentration in water (mg L^{-1}); AE = assimilation efficiency from ingested particulate matter (%); IR = ingestion rate of the matter (g g^{-1} day^{-1}); C_f = metal concentration in food (μg g^{-1}); k_e = efflux rate constant (day^{-1}); and k_g = growth rate constant (day^{-1}).

The dietary component of (1) was re-arranged to calculate the TTF as follows (Wang and Fisher 1999):

$$\text{TTF} = \frac{\text{AE} \times \text{IR}}{k_e + k_g}. \tag{2}$$

The assumption of steady state for some of the metals evaluated may be questioned, especially for temporally dynamic detoxification processes like metallothionein induction and granule formation (Vivjer et al. 2004). Mathematically, all organisms will reach steady state, although in some rare instances (e.g., Zn for barnacles) time to steady state may exceed the organism's lifespan. Equation 1 is simply the solution for a series of differential equations describing uptake and loss of metals at any instantaneous point in time. The general form of the equation used to estimate trophic transfer is valid whether solved for steady state or as an instantaneous measurement. Indeed, the measurements made to estimate TTF for the biokinetic data are short term and not in steady state. Hence the demonstration of steady state is unnecessary for purposes of this review.

Values reported in the literature for terms in (2) were used to generate species-specific TTFs. When unavailable, values for the same species that had been reported in another study (often by the same laboratory) were substituted. In a few cases, in which species-specific IR data were unavailable, a mean IR for the taxonomic group (e.g., bivalves, copepods) was substituted. Taking this step probably had minor effects on biokinetic TTF estimates, because IR generally is not a controlling parameter of TTF and the variability in IR among taxonomic groups generally is low (Luoma and Rainbow 2005; Wang and Rainbow 2008). Similarly k_g was reported infrequently, but with few exceptions does not greatly influence TTF magnitude (Luoma and Rainbow 2005; Wang and Rainbow 2008). Finally, most of the terms in (1) and (2) (k_u, AE, IR, k_e, k_g) are variable functions rather than constants, because exposure concentration influences k_u, k_e, and AE; the organism's exposure history influences k_u, k_e, and AE; and food density influences IR. No attempt was made to normalize these functions to a constant condition because, in most cases, the data needed to parameterize the functions were unavailable. As a result, some of the variations within and among species will be attributable to these sources.

Upon compilation of the biokinetic TTF estimates, species were assigned subjectively to trophic levels and food chains. The food chain assignments were dictated by the species for which biokinetic data were available. Initially, data were pooled into different taxonomic groups—e.g., copepods, bivalves, trophic level 3 fish and trophic level 4 fish. Within a trophic level, the mean ($\pm$ standard error of the mean if n was $\geq$3) of available $\log_{10}$-transformed TTFs was quantified. This allowed for evaluation of specific food chain types (e.g., benthic vs. pelagic). Where both biomagnification potential and variation within taxonomic group TTF were high, the data for individual species were examined to identify possible reasons for the variability. Freshwater and marine biokinetic data were compiled and are presented in Online Resources 2 and 3.

2.3   Field Data

Field bioaccumulation data were compiled for lakes, streams, estuaries, and coastal marine waters (see Online Resource 4). Studies typically comprised food chains of varying lengths and trophic levels, including primary producers, secondary consumers, and predators. We focused generally on studies in which trophic levels were assigned through measurement of stable isotope ratios of nitrogen (^{15}N:^{14}N; $\delta^{15}N$) and carbon (^{13}C:^{12}C; $\delta^{13}C$). As summarized in Croteau et al. (2005), $\delta^{15}N$ is a tool for inferring the relative trophic position of an individual species in a food web, and $\delta^{13}C$ can be used to identify food sources provided they have distinctive isotopic signatures. Food web associations also were inferred from reported dietary preferences of the species (e.g., Timmermans et al. 1989; Barwick and Maher 2003), and when trophic levels could be surmised (e.g., primary producer, primary consumer, secondary consumer). These choices augmented the database with information from a wider variety of sites and food webs. Where possible, linear regression was used to assess slopes and the statistical significance of the relationships between log whole body tissue concentration and trophic levels and/or stable isotope ratios (Online Resource 4). Otherwise, the statistics referenced are those reported in the studies.

3   Results

3.1   Laboratory Data

Laboratory-based TTFs were highly variable for each metal, ranging over two to three orders of magnitude (Online Resource 1). The TTFs for Cd, Cu, Pb, and Zn decreased with increased dietary exposure concentration (the slopes of the relationships between TTFs and dietary exposure concentration were significant, $p \leq 0.003$). TTF variability in relation to whole body concentration was greatest for Ni, possibly reflecting limited data, low TTF magnitude (TTFs generally ≤ 0.1) and differences among taxa (Fig. 1a–e). Metal-specific regressions for taxonomic groups (i.e., arthropod, fish, annelid, etc.) were typically less variable, and statistically significant ($p < 0.05$) for Cd, Cu, Pb, and Zn, indicating that the overall inverse relationship did not result from pooling data from different taxonomic groups.

Ranges of minimal nutritional requirements for Cu and Zn for fish and some invertebrates (Brown 2005) are shown in Fig. 1b, e. The TTF data for Cu and Zn in fish extend above and below the ranges. These data suggest that some of the Cu and Zn TTFs >1 were associated with nutritional requirements.

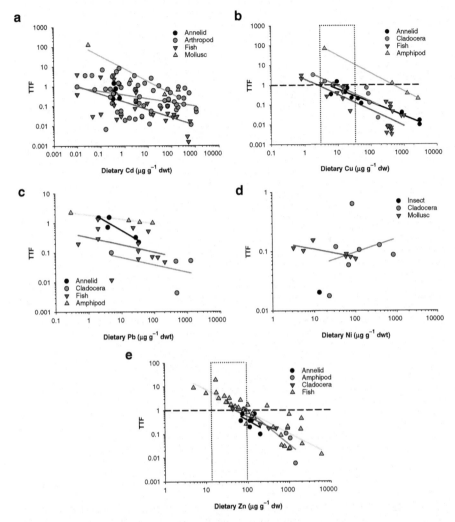

Fig. 1 Relationship between trophic transfer factors for different taxonomic groups and dietary exposure concentration from laboratory studies: (**a**) cadmium, (**b**) copper, (**c**) lead, (**d**) nickel, and (**e**) zinc. The *horizontal dashed line* indicates a TTF of 1. *Solid lines* represent regressions of TTF vs. dietary metal for individual taxa. The *dotted boxes* in (**b**) and (**e**) identify the range nutritional requirements for fish and some invertebrates (from Brown 2005)

Dietary metal toxicity was not more likely when the consumer had bioaccumulated a higher metal concentration than that in its food (Fig. 2a–e; Online Resource 1). Toxicity was defined as a significant adverse effect (i.e., reduced survival, growth, or reproduction) relative to the control. Toxicity based on whole body concentrations appeared unrelated to whether the organism bioaccumulated metal concentrations higher than those in its food, because there were insignificant relationships ($p = 0.2$–0.7) between TTF magnitude and toxicity for Cd, Cu, Pb, and Zn (Fig. 2a–c, e). Almost equal proportions of Cd TTFs >1 was associated with each of the three

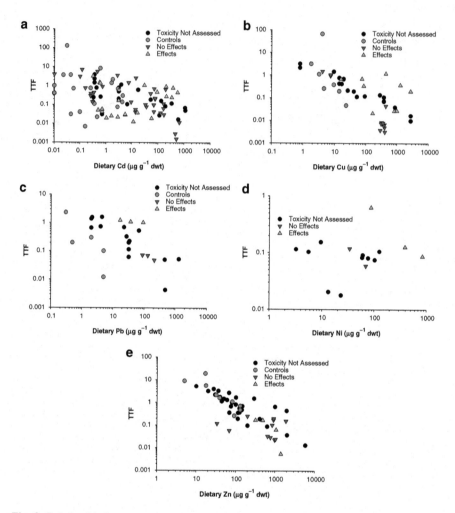

Fig. 2 Relationship between presence–absence of adverse effects, trophic transfer factors and dietary exposure concentration from laboratory studies: (**a**) cadmium, (**b**) copper, (**c**) lead, (**d**) nickel, and (**e**) zinc. The *different symbols* distinguish data associated with effects or lack thereof, and include controls and bioaccumulation studies in which toxicity was not evaluated explicitly, but occurred nonetheless

groups—controls, no effect, and effects (Fig. 2a). For Cu, only one of the nine data points associated with effects had a TTF >1.0 (Fig. 2b). All three data points associated with Pb toxicity had TTFs of 1.0–1.2 (Fig. 2c), all from a study in which the freshwater amphipod *Hyalella azteca* was fed rabbit chow equilibrated with aqueous Pb (Besser et al. 2005). Approximately 75% of the remaining Pb TTF data points in Fig. 2c were <1. None of the Ni TTFs exceeded 1.0 (Fig. 2d), and all of the Zn TTFs >1.0 were associated with control organisms or with studies in which toxicity was not evaluated (Fig. 2e). TTFs were well below 1.0 for the few studies in which dietary Zn toxicity was observed.

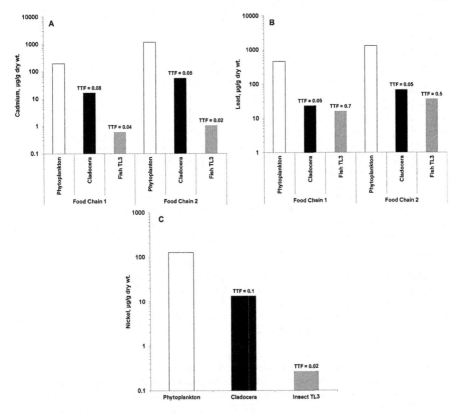

Fig. 3 Trophic transfer data from laboratory-simulated food chain studies for freshwater systems. (**a**) Cadmium, (**b**) lead, and (**c**) nickel. Geomean value for each taxonomic group. *White bars* = metal conc. in phytoplankton, *Black bars* = metal conc. in TL2, *Light gray bars* = metal conc. in TL3. Cadmium data from Ruangsomboon and Wongrat (2006); lead data from Vighi (1981); and nickel data from Ponton and Hare (2010)

Most (93%) of the TTFs summarized in Figs. 1a–e and 2a–e were from studies in which a single trophic transfer was evaluated. In the remaining studies, two or more transfers were evaluated, providing a more complete picture of trophic transfer and facilitating comparison of results to those from biokinetic modeling and field studies. In freshwater, phytoplankton were exposed to aqueous concentrations of Cd, Pb, and Ni and were then fed to cladocerans, which in turn were fed to fish or a predatory insect. Tissue concentrations decreased with increasing trophic level (Fig. 3a–c). From trophic level 1 (TL1) to TL2, the TTFs were similar (0.05–0.1) for Cd, Pb, and Ni. From TL2 to TL3, they declined for Cd (0.07–0.03), declined for Ni (0.1–0.02), and remained unchanged for Pb (0.05–0.6). Results of the single marine study that used Cd were similar to those in freshwater, namely TTFs of 0.1 for the first trophic transfer and one of 0.03 for the next transfer (Fig. OR1-1 in Online Resource 1).

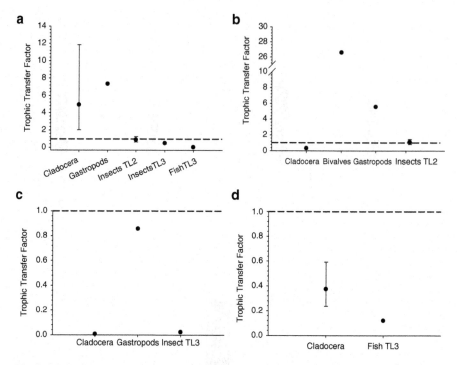

Fig. 4 Trophic transfer factor estimates from biokinetic data for freshwater studies. (**a**) Cadmium, (**b**) copper, (**c**) nickel, and (**d**) zinc. Geomean value for each taxonomic group, except when $n \geq 3$, geometric mean ± standard error of the mean. *Dashed lines* indicate TTF = 1

TTFs from one-step trophic transfers were also applied to simulated food chains because laboratory data for three or more trophic levels were limited ($n = 4$; Figs. OR1-2 and OR1-3 in Online Resource 1). Using these data, TTFs also tended to decline with increasing trophic level for Cd, Cu, and Ni. However, zinc did not: declining slightly in freshwater food chains (ANOVA, $p = 0.07$, $n = 5$ and 9 for TL2 and TL3, respectively) and increasing slightly in the marine food chain (ANOVA, $p = 0.37$, $n = 4$, 7, and 8 for TL2, TL3, and TL4, respectively). Comparatively high TTFs were almost always associated with low dietary exposure concentrations (Fig. 1). Moreover, TTFs >1 for essential metals were associated with organisms assimilating metal to meet minimal nutritional requirements (Fig. 1b, e; Brown 2005).

3.2 Biokinetic Data

Overall, the biokinetic data for metal exposures in freshwater were much more limited than those in saltwater (Online Resource 2). With few exceptions, only a single TTF estimate was available for any taxonomic group, and only three trophic levels could be assessed (Cd, Ni, and Zn) (Fig. 4a, c, d). Copper TTF estimates were only

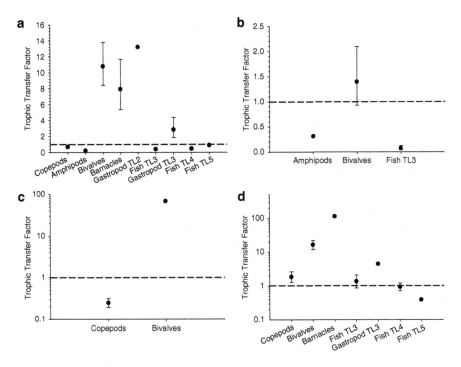

Fig. 5 Trophic transfer factor estimates from biokinetic data for marine studies. (**a**) Cadmium, (**b**) copper, (**c**) nickel, and (**d**) zinc. Geometric mean value given for each taxonomic group, except when $n \geq 3$, geometric mean ± standard error of the mean. *Dashed lines* indicate TTF = 1

available for TL1 to TL2 (Fig. 4b). Given the limited data, it is not possible to draw conclusions regarding the potential for metal biomagnification based on biokinetic data alone for freshwater organisms. The data suggest TTFs >1 from TL1 to TL2 for Cd and Cu, but not for Ni and Zn (Fig. 4a–e). Cadmium TTFs were >1 for both cladocerans and gastropods. Copper TTFs were low (0.1) for cladocera and high (6–27) for gastropods and bivalves. These observations appear to be consistent with the understanding that aquatic organisms in general do not regulate Cd, while crustaceans regulate Cu and mollusks generally do not (Rainbow 1997a, b).

In marine systems, the biokinetic data allowed a more robust assessment of biomagnification potential (Online Resource 3). Up to five trophic levels, multiple taxonomic groups, and multiple species within groups were examined. Trophic transfer factors for Cd and Zn suggest that the potential exists for biomagnification in carnivorous gastropods at both TL2 and TL3, when they prey upon bivalves, barnacles, or herbivorous gastropods in three-step food chains (Fig. 5a, d). Generally, this was not the case for TL3 fish, for which mean TTFs were <1 for Cd and Cu and averaged 2.2 for Zn (Fig. 5 a, b, and d). At TL4 (fish) and TL5 (shark), TTFs were ≤1 for both Cd and Zn. Biokinetic data for Ni were only available for TL1 to TL2, so no conclusions could be drawn (Fig. 5c). The high TTFs for Ni (9.3 and 496) estimated in the single study of Ni in bivalves contrasted with those (<1) for TL2 copepods fed algae, allowing no general conclusion.

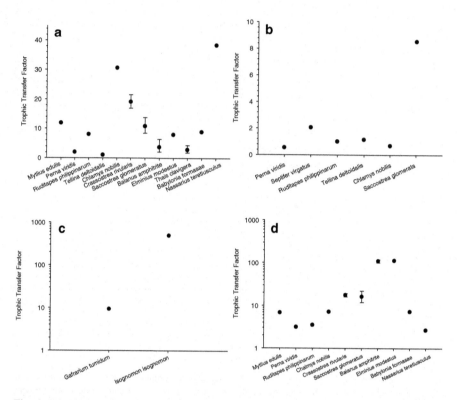

Fig. 6 Trophic transfer factor estimates from biokinetic data for accumulator species in marine systems. (**a**) Cadmium, (**b**) copper, (**c**) nickel, and (**d**) zinc. Geometric mean value for each species. When $n \geq 3$, mean ± SEM

Because of the potential importance of metal biomagnification in barnacles and mollusks in marine systems, we estimated TTFs for these taxa using biokinetic data (Fig. 6a–d). Bivalve TTFs consistently were higher in oysters and scallops (Ostreoida and Pterioida) than in mussels (Mytiloida) and clams (Veneroida) for all metals: Cd, Cu, Ni, and Zn. Both barnacle species had similar TTFs, perhaps not surprising because they are both within the family Balanidae. There were differences in Cd TTFs for the three carnivorous gastropods, with *Thais clavigera* having a distinctly lower mean TTF (1.5) than *Babylonia formosae* (TTF = 8.9) and *Nassarius teretius-culus* (TTF = 38).

3.3 Field Data

The majority of authors performing field studies (75% of 25 papers) that addressed metal concentrations at multiple trophic levels of aquatic organisms reported little evidence for the biomagnification of Cd, Cu, Pb, Ni, or Zn among primary producers, primary consumers, and predators, including fish (Online Resource 4). Studies in which metal biomagnification was reported were generally limited to specific macroinvertebrate food webs; in fact, only two studies, both involving Zn, provided evidence of biomagnification in fish.

Several studies, in which trophic levels were defined using stable isotopes, provided evidence of significant ($p \leq 0.05$) biomagnification of certain metals in macroinvertebrates, but not fish (Online Resource 4). Examples of Cd biomagnification were limited to invertebrate food chains, and one involving a predatory flatworm (*Dugesia tigrina*) had a mean Cd concentration approximately three times greater than other macroinvertebrates within the same food web (Croteau et al. 2005). Another example involved herbivorous mollusks and predacious mollusks occupying TL 2 and 3 (Cheung and Wang 2008), whose results were consistent with the biokinetic estimates presented above. The best evidence of biomagnification in this last study involved Cd in two of the three habitats sampled, namely rocky intertidal habitats. Analysis of the $\delta^{15}N$-$\log_{10}$ Cd data that were reported, which were based on all species sampled in each intertidal habitat, indicated positive slopes ($b = 0.16$–0.24) that were statistically significant ($p = 0.002$–0.011) for both bays. For Cu and Zn, Jara-Marini et al. (2009) observed a significant ($p < 0.05$) linear relationship between Cu concentrations and $\delta^{15}N$ values from phytoplankton up to a secondary consumer crab, but the relationship was insignificant ($p > 0.05$) when secondary and tertiary consumer fish were included. The highest Zn concentrations were several-fold greater in filter-feeding oysters and barnacles than in other higher trophic level macroinvertebrates, such as snails, polychaetes, shrimp, and crab. Field data from Timmermans et al. (1989) and Quinn et al. (2003) both suggested that Zn biomagnified—but not Cd, Cu, and Pb—in insects occupying multiple trophic levels. The conclusion of Quinn et al. (2003) concerning zinc biomagnification might deserve further qualification as, based on re-analysis of their data, the slopes of the TMF regressions were significant only for one of the four stream/year combinations studied. Zinc biomagnification [TMF = 0.43 (95%CL = 0.3–0.6)] was only significant in insects from the uncontaminated stream in 1999 and was not significant ($p = 0.13$) the next year, 2000 [TMF = 0.09 (95%CL = −0.02 to 0.2)]. Biomagnification did not occur in insects from the stream receiving acid mine drainage in either year.

Other studies have failed to confirm Zn biomagnification in the field. Farag et al. (1998), for example, observed significantly ($p \leq 0.01$) lower Zn concentrations in detritivorous, omnivorous and carnivorous benthic macroinvertebrates from the metals-contaminated Coeur d'Alene River (Idaho) than in herbivorous benthic macroinvertebrates. Similarly, Watanabe et al. (2008) found insignificant correlations ($p > 0.05$) among Zn concentrations and $\delta^{15}N$ values in macroinvertebrates collected from three of four locations in a creek below an abandoned mining area in Japan.

There were two field studies in which Zn concentrations in fish were greater than in lower trophic levels. However, it appears that these higher Zn concentrations in fish are due to essentiality requirements, because the Zn concentrations in the food web were relatively low. In the first study, Saiki et al. (1995) analyzed Zn in several food web organisms collected from the mining-impacted locations in the upper Sacramento River (USA) and in reference tributaries. In reference tributaries, mean

Zn concentrations in fish were similar to or greater than those measured in macro-invertebrates and the aquatic plant *Elodea canadensis* (i.e., TTFs > 1.0) At the mining-impacted locations, however, mean Zn concentrations in fish were always less than the mean concentrations in macroinvertebrates and waterweed (i.e., TTFs < 1.0). The increased Zn concentration in fish relative to lower trophic levels collected from reference streams, but not in mining-impacted streams, likely reflects internal regulation of Zn by the fish. In the second study (Campbell et al. 2005), Zn concentrations in Arctic cod (*Boreogadus saida*) collected from northern Baffin Bay (between Ellesmere Island, Canada, and Greenland) were greater than in the food web. Because this study area is not contaminated with metals, the higher Zn residues in the cod probably reflect physiological requirements.

4   Conclusions and Discussion

Our review of the freshwater and saltwater literature concerning trophic transfer and biomagnification suggested several conclusions that we summarize below.

4.1   Overall Findings Concerning Biomagnification Potential

Cadmium, Cu, Ni, Pb, and Zn generally do not biomagnify in food chains consisting of primary producers, macroinvertebrate consumers and fish. Yet, there are specific food chains in which biomagnification of these metals occurs. First, biomagnifica-tion may occur in certain marine food chains consisting of bivalves, herbivorous gastropods, and barnacles at TL 2 and carnivorous gastropods at TL3. Trophic transfer factors for all three groups of taxa at TL2 were consistently >1 for these four metals (Online Resource 3). Moreover, their gastropod predators at TL3 had Cd and Zn TTFs significantly >1. Depending on the subcellular metal distribution in these gastropods, the possibility of direct toxicity to the snails at TL3 or to their predators at TL4 warrants further study.

Secondly, biomagnification of Zn in fish (TTFs of 1–2) may occur in specific circumstances. The authors of four biokinetic studies reported TTFs of 1.1–6.0 for fish feeding on planktonic crustacean and clams, and, in one, a TTF of 0.2 (Online Resource 3). Yet all but one laboratory-derived TTF >1, for a TL3 fish, was associ-ated with dietary Zn concentrations ≤100 µg/g dry wt, the upper end of the essential range for dietary Zn (Fig. 1e). Thus, it is possible that the test fish may have been deficient in Zn. In addition, some of the values used to parameterize the biokinetic model may be suspect: The highest Zn TTFs for fish of 4.9 and 6.0 did not quantify K_g, which can affect biokinetic results because it can be comparable to efflux rate (Baines et al. 2002; Dutton and Fisher 2010). This may have produced an overesti-mate of the TTF (Xu and Wang 2002; Zhang and Wang 2007). Overall, it appears that Zn TTFs between 1 and 2 are possible for fish.

In field studies with fish, Zn biomagnification only occurred in uncontaminated reference sites where prey contained generally ≤105 µg Zn/g dry wt (Campbell et al. 2005; Saiki et al. 1995). Zn biomagnification was neither observed in reference streams where macroinvertebrate prey had 235–977 µg Zn/g dry wt (Farag et al. 2007), nor in mining-impacted streams with macroinvertebrate Zn concentrations ranging from 430 to 1,600 µg/g dry wt (Saiki et al. 1995). Overall, the weight of evidence suggests Zn biomagnification is more likely to occur in waters where ambient Zn concentrations are deficient or less than optimal.

4.2 Congruence of Lab, Field, and Modeled TTF Estimates

We assessed trophic transfer in aquatic systems using three independent methodologies: laboratory food chain, biokinetic modeling and field studies. In general they yielded similar conclusions. The one exception to method congruence was the relationship between lab and modeled TTFs for freshwater cladocerans exposed to dietary Cd. The majority (79%, $n = 24$) of the lab Cd TTFs for cladocerans were <1. Modeled TTFs for *Daphnia* ranged from 1.3 to 4.7, except for a TTF of 0.21 that was based on a lower ingestion rate and higher elimination rate than the other daphnid studies. Despite the uncertainty in Cd TTFs for cladocerans, all three methodologies indicate that the TTFs for organisms (*Chaoborus* and two species of fish) consuming cladocerans are always <1, thus signifying no biomagnification.

4.3 The Inverse Relationship Between TTF and Metal Concentrations

There was an inverse relationship between TTF and exposure concentration for Cd, Cu, Pb, and Zn. This finding is consistent with previous reviews of bioconcentration factors, bioaccumulation factors, and TTFs for metals (DeForest et al. 2007; McGeer et al. 2003). Moreover, the inverse relationships were most pronounced for the essential metals Cu and Zn and more variable for the non-essential metals Cd and Pb. The underlying mechanisms for the inverse relationship vary with metal and species, reflecting variations among species in metal regulation and subcellular storage strategies (Phillips and Rainbow 1989; Rainbow and White 1989; Rainbow et al. 1990). For essential metals like Cu and Zn, internal tissue concentrations may be regulated over a fairly wide range of exposure concentrations, with TTFs decreasing as dietary exposure increases. For aquatic organisms that detoxify metals by storage (e.g., granules), TTFs increase with increasing dietary exposure, but not necessarily proportionally. This phenomenon appears to be partially explained by decreases in assimilation efficiencies as exposure concentration increases (e.g., Reinfelder et al. 1998; Guan and Wang 2004; Croteau and Luoma 2008; Lapointe et al. 2009). The effect on assimilation efficiency may result from Michaelis–Menten type saturation kinetics for metal transport proteins in the digestive epithelia (Bury et al. 2003).

4.4   The Relationship Between TTFs and Toxicity

Our analysis also failed to demonstrate a relationship between the magnitude of TTFs and dietary toxicity to consumers/predators. Consequently, TTFs for the metals examined may not be an inherently useful predictor of potential hazard (i.e., toxic potential) to aquatic organisms. Perhaps, as our scientific understanding develops, it will be possible to use other metrics, such as the biologically effective dose, to estimate effects of assimilated metal. This is the not possible presently because of many unknowns, such as the fraction of metal that is bioavailable to organisms in the next trophic level, which is highly species-specific. For example, fish cannot digest metal granules sequestered by bivalves but predatory snails can. The biologically effective dose has not been defined for metals and, based on the limited data available, is unlikely to be consistent across taxa (Adams et al. 2011).

4.5   Extrapolation of Results to the Field

Finally, we note that the laboratory and biokinetic studies reported above have focused on highly controlled studies of structured, simple food chains, where the predator has no choice of prey type. They also have tended to focus on trophic levels 1–3 and mainly invertebrates. Many aquatic food webs in nature encompass not only more trophic levels but are unstructured (Isaacs 1973), wherein species feed at multiple trophic levels as prey availability varies over time and space as a result of changes in prey body size, energy content, densities, etc. (Petchey et al. 2008). These complexities raise uncertainties about the degree to which lab and biokinetic modeling results can be extrapolated to the field. Yet, this meta-analysis indicated congruence between the lab and field studies and the biokinetic modeling. The field results are especially compelling. For example, the authors of one field study of an unstructured food web off the Southern California coast in the 1970s failed to detect biomagnification of Cd, Cu, and Zn in food webs encompassing five trophic levels from zooplankton to Great White Shark (Schafer et al. 1982) (Fig. 7a). Yet clear-cut evidence of organic mercury's biomagnification was observed in the same food web (Fig. 7b). Their results support this paper's inferences regarding the lack of biomagnification of Cd, Cu, Ni, Pb, and Zn in most aquatic food chains.

4.6   Data Variability and Additional Uncertainties

Most of the studies of biomagnification have not accounted for variability within species, over time or between habitats. In addition, biokinetic data were limited in some cases, precluding conclusions concerning the potential for metal biomagnification solely from these data. Data limitations were especially notable for Ni, although 91% ($n=23$) of the TTFs for this metal compiled in Online Resources 1, 2, and 3 were <1.0.

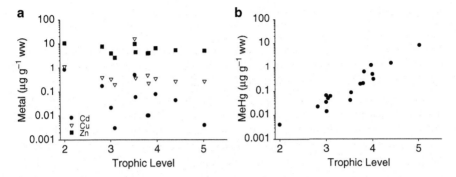

Fig. 7 Tissue residues (mostly muscle) of (**a**) Cd, Cu, and Zn and (**b**) methylmercury, in relation to trophic level in 11 species (9 fish, zooplankton, and squid) from a coastal pelagic food chain off Southern California. Data from Table 2 of Schafer et al. (1982)

Another set of uncertainties involve how background tissue concentrations for essential metals influence the results of field studies, and whether it is possible to distinguish true increases in metal residues due to trophic transfer from inherent differences between species. Unfortunately, background metal concentrations in tissues are rarely measured. This hinders interpretation of data from lab and field studies, but not biokinetic studies that use radio-isotopes, where TTFs are based on accumulation and efflux of new metal, and therefore are not subject to interference from background metal concentrations.

In the field, more variability can be expected because the food webs are more temporally and spatially dynamic, and unstructured. Sampling food webs is difficult because spot sampling reflect what species are present and possess sufficient mass to accommodate the analyses sought. Moreover, variability among samples of the same specimens is rarely integrated into regressions of $\delta^{15}N$ or TL vs. log metal residue. Sampling location, variation between years and variation in species composition have the potential to materially influence results. For example, in a study of Cu, Fe, and Zn in insects occupying TL 2–3, the relationship between log metal residue and TL (or $\delta^{15}N$) differed between years and locations (control vs. mine drainage) sampled (Quinn et al. 2003). The data of Cheung and Wang (2008) were similar: Cd biomagnified in invertebrates collected at two rocky intertidal habitats, but not in a soft-bottomed, subtidal habitat. Metal residues within and between taxa vary widely, of course, and means or values from single samples typically constitute the dependent variable in the TL-log metal regressions. For example, such variation, indexed in terms of coefficient of variations, ranged from about 14–69% for Cd in specimens from Butterfly and Clearwater bays. Yet the independent variable, TL or $\delta^{15}N$, which in typical regression is assumed to be measured without error, does vary: coefficients of variation ranging from 1 to 6% for $\delta^{15}N$ measurements in the Cheung and Wang (2008) study and from 1 to 16% in the Jara-Marini et al. (2009) study. Despite these sources of error, many of them were not measured or integrated into the studies. Yet, overall, the congruence of the multiple lines of evidence increases confidence in the conclusions of this assessment.

5 Summary

In this review, we sought to assess from a study of the literature whether five inorganic metals (viz., cadmium, copper, lead, nickel, and zinc) biomagnify in aquatic food webs. We also examined whether accumulated metals were toxic to consumers/predators and whether the essential metals (Cu and Zn and possibly Ni) behaved differently from non-essential ones (Cd and Pb). Biomagnification potential was indexed by the magnitude of single and multiple trophic transfers in food chains. In this analysis, we used three lines of evidence—laboratory empirical, biokinetic modeling, and field studies—to make assessments. Trophic transfer factors, calculated from lab studies, field studies, and biokinetic modeling, were generally congruent. Results indicated that Cd, Cu, Pb, and Zn generally do not biomagnify in food chains consisting of primary producers, macroinvertebrate consumers, and fish occupying TL 3 and higher. However, biomagnification of Zn (TTFs of 1–2) is possible for circumstances in which dietary Zn concentrations are below those required for metabolism. Cd, Cu, Ni, and Zn may biomagnify in specific marine food chains consisting of bivalves, herbivorous gastropods, and barnacles at TL2 and carnivorous gastropods at TL3. There was an inverse relationship between TTF and exposure concentration for Cd, Cu, Pb, and Zn, a finding that is consistent with previous reviews of bioconcentration factors and bioaccumulation factors for metals. Our analysis also failed to demonstrate a relationship between the magnitude of TTFs and dietary toxicity to consumer organisms. Consequently, we conclude that TTFs for the metals examined are not an inherently useful predictor of potential hazard (i.e., toxic potential) to aquatic organisms. This review identified several uncertainties or data gaps, such as the relatively limited data available for nickel, reliance upon highly structured food chains in laboratory studies compared to the unstructured food webs found in nature, and variability in TTFs between the organisms found in different habitats, and years sampled.

Acknowledgments This assessment was partially funded by Rio Tinto. KVB was supported by the SETAC/ICA Chris Lee Award for Metals Research. This manuscript was improved based on the reviews performed by Drs. Jim Meador and Alan Mearns, and the Journal editor. We are grateful for their efforts.

References

Adams WJ, Blust R, Borgmann U, Brix KV et al (2011) Utility of tissue residues for predicting effects of metals on aquatic organisms. Integr Environ Assess Manage 7:75–98

Baines SB, Fisher NS, Stewart AR (2002) Assimilation and retention of selenium and other trace elements from crustacean food by juvenile striped bass (*Morone saxatilis*). Limnol Oceanogr 47(3):646–655

Baptist JP, Lewis CW (1967) Transfer of ^{65}Zn and ^{51}Cr through an estuarine food chain. In: Nelson DJ, Evans FC (eds) Symposium on radioecology, proceedings of the second national symposium. U.S. Atomic Energy Commission, Ann Arbor, MI, pp 420–430

Barwick M, Maher W (2003) Biotransference and biomagnification of selenium, copper, cadmium, zinc, arsenic and lead in a temperate seagrass ecosystem from Lake Macquarie Estuary, NSW, Australia. Mar Environ Res 56:471–502

Baudin J, Nucho R (1992) ^{60}Co accumulation from sediment and planktonic algae by midge larvae (*Chironomus luridus*). Environ Pollut 76:133–140

Besser JM, Brumbaugh WG, Brunson EL, Ingersoll CG (2005) Acute and chronic toxicity of lead in water and diet to the amphipod *Hyalella azteca*. Environ Toxicol Chem 24:1807–1815

Biddinger GR, Gloss SP (1984) The importance of trophic transfer in the bioaccumulation of chemical contaminants in aquatic ecosystems. Res Rev 91:103–145

Borga K, Kidd KA, Muir DCG, Berglund O, Conder JM, Gobas FAPC (2011) Trophic magnification factors: considerations of ecology, ecosystems, and study design. Integr Environ Assess Manage 8:64–84

Boroughs H, Chipman WA, Rice TR (1957) Laboratory experiments on the uptake, accumulation, and loss of radionuclides by marine organisms (Chapter 8). In: National Academy of Sciences (ed) The effects of atomic radiation on oceanography and fisheries (Publication 551). National Academy of Sciences, National Research Council, Washington, DC

Brown PB (2005) Nutritional considerations for fish and invertebrates. In: Meyer JS, Adams WJ, Brix KV, Luoma SN, Mount DR, Stubblefield WA, Wood CM (eds) Toxicity of dietborne metals to aquatic organisms. SETAC Press, Pensacola, FL, 329 pp

Bruggeman WA, Operhuizen A, Wibenga A, Hutzinger O (1984) Bioaccumulation of super-lipophilic chemicals in fish. Toxicol Environ Chem 7:173–189

Burdick G, Harris E, Dean H, Walker T, Skea J, Colby D (1964) The accumulation of DDT in lake trout and the effect on reproduction. Trans Am Fish Soc 93:127–136

Bury NR, Walker PA, Glover CN (2003) Nutritive metal uptake in teleost fish. J Exp Biol 206:11–23

Campbell LM, Norstrom RJ, Hobson KA, Muir DCG, Backus S, Fisk AT (2005) Mercury and other trace elements in a pelagic Arctic marine food web (Northwater Polynya, Baffin Bay). Sci Total Environ 351–352:247–263

Cheung MS, Fok EMW, Ng TYT, Yen YF, Wang WX (2006) Subcellular cadmium distribution, accumulation, and toxicity in a predatory gastropod, *Thais clavigera*, fed different prey. Environ Toxicol Chem 25:174–181

Cheung MS, Wang WX (2008) Analyzing biomagnification of metals in different food webs using nitrogen isotopes. Mar Pollut Bull 56:2082–2105

Croteau MN, Luoma SN, Stewart AR (2005) Trophic transfer of metals along freshwater food webs: evidence of cadmium biomagnification in nature. Limnol Oceanogr 50:1511–1519

Croteau MN, Luoma SN (2008) A biodynamic understanding of dietborne metal uptake by a freshwater invertebrate. Environ Sci Technol 42:1801–1806

Croteau MN, Luoma SN (2009) Predicting dietborne metal toxicity from metal influxes. Environ Sci Technol 43:4915–4921

Dallinger R, Prosi F, Segner H, Back H (1987) Contaminated food and uptake of heavy metals by fish: a review and a proposal for further research. Oecologia 73:91–98

Davis J, Foster R (1958) Bioaccumulation of radioisotopes through aquatic food chains. Ecology 39:530–535

Davis DA, Gatlin M III (1996) Dietary mineral requirements of fish and marine crustaceans. Rev Fish Sci 4(1):75–99

DeForest DK, Brix KV, Adams WJ (2007) Assessing metal bioaccumulation in aquatic environments: the inverse relationship between bioaccumulation factors, trophic transfer factors and exposure concentration. Aquat Toxicol 84:236–246

Dillon TM, Suedel BC, Peddicord, RK, Clifford, PA, Boraczek JA (1995) Trophic transfer and biomagnification potential of contaminants in aquatic ecosystems. EEDP-01-033. Environmental Effects of Dredging Technical Notes. US Army Engineer Waterways Experiment Station, Vicksburg, Miss 12 pp

Dutton S, Fisher NS (2010) Intraspecific comparisons of metal bioaccumulation in the juvenile Atlantic silverside *Menidia menidia*. Aquat Biol 10:211–226

Farag AM, Woodward DF, Goldstein JN, Brumbaugh W, Meyer JS (1998) Concentrations of metals associated with mining waste in sediments, biofilm, benthic macroinvertebrates, and fish from the Coeur d'Alene River basin. Idaho. Arch Environ Contam Toxicol 34:119–127

Farag AM, Nimick DA, Kimball BA, Church SE, Harper DD, Brumbaugh WG (2007) Concentrations of metals in water, sediment, biofilm, benthic macroinvertebrates, and fish in the Boulder River watershed, Montana, and the role of colloids in metal uptake. Arch Environ Contam Toxicol 52:397–409

Fisk AT, Norstrom RJ, Cymbalisty CD, Muir CDG (1998) Dietary accumulation and depuration of hydrophobic organochlorines: bioaccumulation parameters and their relationship with the octanol/water partition coefficient. Environ Toxicol Chem 17:951–961

Garnier-Laplace J, Vray F, Baudin J (1997) A dynamic model for radionuclide transfer from water to freshwater fish. Water Air Soil Pollut 98:141–166

Guan R, Wang W-X (2004) Dietary assimilation and elimination of Cd, Se, and Zn by *Daphnia magna* at different metal concentrations. Environ Toxicol Chem 23:2689–2698

Hunt EG (1966) Biological magnification of pesticides. In: National Research Council (ed) Scientific aspects of pest control (Publication 1402). National Academy of Sciences and National Research Council, Washington, DC

Isaacs JD (1973) Unstructured marine food webs and pollutant analogues. Fish Bull 70:1053–1059

Jara-Marini ME, Soto-Jiménez MF, Páez-Osuna F (2009) Trophic relationships and transference of cadmium, copper, lead and zinc in a subtropical coastal lagoon food web from SE Gulf of California. Chemosphere 77:1366–1373

Ju Y-R, Chen W-Y, Singh S, Liao C-M (2011) Trade-offs between elimination and detoxification in rainbow trout and common bivalve molluscs exposed to metal stressors. Chemosphere 85:1048–1056

Krumholz LA, Foster RF (1957) Ecological factors involved in the uptake, accumulation, and loss of radionuclides by aquatic organisms (Chapter 7). In: National Academy of Sciences (ed) The effects of atomic radiation on oceanography and fisheries (Publication 551). National Academy of Sciences, National Research Council, Washington, DC

Lapointe D, Gentes S, Ponton DE, Hare L, Couture P (2009) Influence of prey type on nickel and thallium assimilation, subcellular distribution and effects in juvenile fathead minnows (*Pimephales promelas*). Environ Sci Technol 43:8665–8670

Luoma SN, Rainbow PS (2005) Why is metal bioaccumulation so variable? Biodynamics as a unifying concept. Environ Sci Technol 39:1921–1931

Macek KJ, Petrocelli SR, Sleight BH (1979) Considerations in assessing the potential for, and significance of biomagnification of chemical residues in aquatic food chains. In: Marking LL, Kimerle RA (eds) Aquatic toxicology, STP 667. American Society for Testing and Materials, Philadelphia, PA, pp 251–268

Mathews T, Fisher NS (2008) Evaluating the trophic transfer of cadmium, polonium, and methylmercury in an estuarine food chain. Environ Toxicol Chem 27:1093–1101

McAlpine D, Araki S (1959) Minamata disease: late effects of an unusual neurological disorder caused by contaminated fish. AMA Arch Neurol 1:522–530

McGeer J, Brix KV, Skeaff JM, DeForest DK, Brigham SI, Adams WJ, Green AS (2003) The inverse relationship between bioconcentration factor and exposure concentration for metals: implications for hazard assessment of metals in the aquatic environment. Environ Toxicol Chem 22:1017–1037

Meador J (2006) Rationale and procedures for using the tissue-residue approach for toxicity assessment and determination of tissue, water, and sediment quality guidelines for aquatic organisms. Human Ecol Risk Assess 12:1018–1073

Peakall DB (1969) Effect of DDT on calcium uptake and vitamin D metabolism in birds. Nature 224:1219–1220

Peakall D, Lincer J (1970) Polychlorinated biphenyls, another long-life widespread chemical in the environment. Bioscience 20:958–964

Petchey OW, Beckerman AP, Riede JO, Warren PH (2008) Size, foraging, and food web structure. Proc Natl Acad Sci U S A 105:4191–4196

Phillips DJH, Rainbow PS (1989) Strategies of trace metal sequestration in aquatic organisms. Mar Environ Res 28:207–210

Ponton DE, Hare L (2010) Nickel dynamics in the lakewater metal biomonitor *Chaoborus*. Aquat Toxicol 96:37–43

Quinn MR, Feng X, Folt CL, Chamberlain CP (2003) Analyzing trophic transfer of metals in stream food webs using nitrogen isotopes. Sci Total Environ 317:73–89

Rainbow PS (1997a) Ecophysiology of trace metal uptake in crustaceans. Estuar Coast Shelf Sci 44:169–175

Rainbow PS (1997b) Trace metal accumulation in marine invertebrates: marine biology or marine chemistry? J Mar Biol Assoc UK 77:195–210

Rainbow PS, White SL (1989) Comparative strategies of heavy metal accumulation by crustaceans: zinc, copper and cadmium in a decapod, an amphipod and a barnacle. Hydrobiologia 174:245–262

Rainbow PS, Phillips DJH, Depledge MH (1990) The significance of trace metal concentrations in marine invertebrates: a need for laboratory investigations of accumulation strategies. Mar Pollut Bull 21:321–324

Rainbow PS, Poirier L, Smith BD, Brix KV, Luoma SN (2006) Trophic transfer of trace metals: subcellular compartmentalization in a polychaete and assimilation by a decapod crustacean. Mar Ecol Prog Ser 308:91–100

Reinfelder JR, Fisher NS, Luoma SN, Nichols JW, Wang W-X (1998) Trace element trophic transfer in aquatic organisms: a critique of the kinetic model approach. Sci Total Environ 219:117–135

Ruangsomboon S, Wongrat L (2006) Bioaccumulation of cadmium in an experimental aquatic food chain involving phytoplankton (*Chlorella vulgaris*), zooplankton (*Moina macrocopa*), and the predatory catfish *Clarias macrocephalus* X C. *gariepinus*. Aquat Toxicol 78:15–20

Saiki MK, Castleberry DT, May TW, Martin BA, Bullard FN (1995) Copper, cadmium, and zinc concentrations in aquatic food chains from the Upper Sacramento River (California) and selected tributaries. Arch Environ Contam Toxicol 29:484–491

Schafer HA, Hershelman, GP, Young, DR, Mearns AJ (1982) Contaminants in ocean food webs. In: Coastal Water Research Project Biennial Report for the Year 1981. 12 pp. [ftp://ftp.sccwrp. org/pub/download/DOCUMENTS/AnnualReports/1981_82AnnualReport/AR81-82_017.pdf]

Schmidt TS, Clements WH, Zuellig RE, Mitchell KA, Church SE, Wanty RB, San Juan CA, Adams M, Lamothe PJ (2011) Critical tissue residue approach linking accumulated metals in aquatic insects to population and community-level effects. Environ Sci Technol 45:7004–7010

Seiler, RL, Skorupa JP (2001) National irrigation water quality program data synthesis. http:// pubs.usgs.gov/of/2000/ofr00513/

Suedel B, Boraczek J, Peddicord R, Clifford P, Dillon T (1994) Trophic transfer and biomagnification potential of contaminants in aquatic ecosystems. Rev Environ Contam Toxicol 136:21–89

Timmermans KR, Van Hattum B, Kraak MHS, Davids C (1989) Trace metals in a littoral foodweb: concentrations in organisms, sediment and water. Sci Total Environ 87(88):477–494

Vighi M (1981) Lead uptake and release in an experimental trophic chain. Ecotoxicol Environ Saf 5:177–193

Vivjer MG, Van Gestel CA, Lanno RP, Van Straalen NM, Peijnenburg WJGM (2004) Internal metal sequestration and its toxicological relevance: a review. Environ Sci Technol 38:4705–4712

Wallace WG, Lopez GR (1997) Bioavailability of biologically sequestered cadmium and the implications of metal detoxification. Mar Ecol Prog Ser 147:149–157

Wallace WG, Lee BG, Luoma SN (2003) Subcellular compartmentalization of Cd and Zn in two bivalves. I. Significance of metal-sensitive fractions (MSF) and biologically detoxified metal (BDM). Mar Ecol Prog Ser 249:183–197

Wang W-X (2002) Interactions of trace metals and different marine food chains. Mar Ecol Prog Ser 243:295–309

Wang WX, Fisher NS (1999) Delineating metal accumulation pathways for marine invertebrates. Sci Total Environ 237–238:459–472

Wang WX, Rainbow PS (2008) Comparative approaches to understand metal bioaccumulation in aquatic animals. Comp Biochem Physiol 148C:315–323

Watanabe K, Monaghan MT, Takemon Y, Omura T (2008) Biodilution of heavy metals in a stream macroinvertebrate food web: evidence from stable isotope analysis. Sci Total Environ 394:57–67

White SL, Rainbow PS (1985) On the metabolic requirements for copper and zinc in molluscs and crustaceans. Mar Environ Res 16:215–229

Xu Y, Wang WX (2002) Exposure and potential food chain transfer factor of Cd, Se and Zn in marine fish *Lutjanus argentimaculatus*. Mar Ecol Prog Ser 238:173–186

Zhang L, Wang W-X (2007) Size-dependence of the potential for metal biomagnification in early life stages of marine fish. Environ Toxicol Chem 26:787–794

Index

Printed in the United States
By Bookmasters